Smart Grid in IoT-Enabled Spaces

Smart Grid in IoT-Enabled Spaces

The Road to Intelligence in Power

Fadi Al-Turjman

CRC Press
Taylor & Francis Group
Boca Raton London New York

CRC Press is an imprint of the
Taylor & Francis Group, an **Informa** business

When you fly high people will throw stones at you. Don't look down. Just fly higher so the stones won't reach you….

Chetan Bhagat

To my wonderful family…
Fadi Al-Turjman

Contents

Preface

Smart grid is a solution, where the Internet of Things (IoT)-enabling technologies have been used towards further advances in energy savings. It tightly integrates with the existing cloud infrastructure to impact several fields in academia and/or industry. IoT-enabled grids have made revolutionary advances in the cloud which we have used so far. It is not only used in computer-assisted energy-aware applications but also in our daily activities towards more intelligent health, traffic, and cyber-physical systems. It comes nowadays enabled with IoT features to integrate smoothly with the emerging 5G networks and beyond in the next-generation telecommunication paradigms. This book opens the door for other exciting research and applications in this era. In this book, we are considering significant research topics towards realizing the future vision of the smart grid in IoT-enabled spaces.

This book overviews these issues and proposes the most up-to-date alternatives. The objective is to pave the way for IoT-enabled spaces in the next-generation smart grid paradigm and open the door for further innovative ideas.

Fadi Al-Turjman

Author

Prof. Dr. Fadi Al-Turjman received his PhD in computer science from Queen's University, Kingston, Canada, in 2011. He is a full professor and a research center director at Near East University, Nicosia, Cyprus. He is a leading authority in the areas of smart/intelligent, wireless, and mobile networks architectures, protocols, deployments, and performance evaluation. His publication history spans over 250 publications in journals, conferences, patents, books, and book chapters, in addition to numerous keynotes and plenary talks at flagship venues. He has authored and edited more than 25 books about cognition, security, and wireless sensor networks' deployments in smart environments published by Taylor & Francis, Elsevier, and Springer. He has received several recognitions and best papers' awards at top international conferences. He also received the prestigious "Best Research Paper Award" from *Elsevier Computer Communications Journal* for the period 2015–2018, in addition to the "Top Researcher Award" for 2018 at Antalya Bilim University, Antalya, Turkey. He has led a number of international symposia and workshops in flagship communication society conferences. Currently, he serves as an associate editor and the lead guest/associate editor for several well-reputed journals including the *IEEE Communications Surveys & Tutorials* (**IF 22.9**) and the *Elsevier Sustainable Cities and Society* (**IF 4.7**).

List of Contributors

Abdulsalam Ahmed Abdulkadir
Engineering Faculty
Middle East Technical University
Guzelyurt, Republic of Northern
  Cyprus

Mohammad Abujubbeh
Engineering Faculty
Middle East Technical University
Guzelyurt, Republic of Northern
  Cyprus

Chadi Altrjman
Engineering Faculty
University of Waterloo
Waterloo, Ontario, Canada

Shehzad Ashraf Chaudhry
Department of Computer Engineering
Faculty of Engineering and
  Architecture
Istanbul Gelisim University
Avcilar, Istanbul, Turkey

B. D. Deebak
Research Center for AI and IoT
Near East University
Nicosia, Mersin, Turkey

and

School of Computer Science and
  Engineering
Vellore Institute of Technology
Vellore, India

Sadia Din
School of Computer Science and
  Engineering
Kyungpook National University
Daegu, Korea

Arman Malekloo
Research Centre for AI and IoT
Near East University
Nicosia, Mersin, Turkey

Ekin Ozer
Department of Computer Engineering
Faculty of Engineering and
  Architecture
Istanbul Gelisim University
Avcilar, Istanbul, Turkey

Anand Paul
School of Computer Science and
  Engineering
Kyungpook National University
Daegu, Korea

Ramiz Salama
Computer Engineering Department

and

Research Centre for AI and IoT
Near East University
Nicosia, Mersin, Turkey

Ramiz Shahroze
Research Centre for AI and IoT
Near East University
Nicosia, Mersin, Turkey

Khalid Yahya
Department of Computer Engineering
Faculty of Engineering and
  Architecture
Istanbul Gelisim University
Avcilar, Istanbul, Turkey

Hadi Zahmatkesh
Department of Computer Engineering
Middle East Technical University
Guzelyurt, Republic of Northern
  Cyprus

1 IoT-Enabled Smart Grid
An Overview

Fadi Al-Turjman
Near East University

Mohammad Abujubbeh
Middle East Technical University

CONTENTS

1.1   INTRODUCTION

Electrical power is one of the essential factors for the development of societies by improving life quality. However, the conditions in power industry are changing as electricity demand and renewable integration are increasing. The stress of increased power demand has produced a burden on the conventional power production resources. With the noticeable decline in conventional power resources' reserves and the recent attention on environmental issues associated with producing power from fossil fuel-based resources, power utilities and investors are motivated to invest into other sustainable ways of power production in order to meet the demand. For instance, one aim of the European 20/20/20 strategy is to increase the share of renewable energy generation up to 20% by 2020 [1]. Renewable resources are intermittent by nature. The increased penetration of those non-dispatchable energy sources into the existing power grid makes it more challenging for utilities to deliver reliable and good-quality power. Fortunately, with the recent technological advancements, IoT-inspired applications can offer great solutions to the aforementioned challenges by providing two-way communication schemes which can help in transforming conventional power grids into modernized SGs. In fact, the IoT paradigm is an essential segment of the modern SGs, especially in residential and commercial buildings' applications [2]. The smart cities paradigm is also another demanding project for IoT-enabled SGs [3]. Both of these emerging paradigms imply that there will be a noticeable increase in the usage of sensor networks for providing useful data that enable the efficient control and management of cities [3,4]. The concept of smart cities will not only focus on specific services such as traffic control but will also extend its means to the electric system. In fact, the usage of sensory devices and SMs in SGs ultimately solves most of electrical industry problems [1,5,6]. In this way, the SG will be able to effectively deal with many aspects of power generation, transmission, and distribution issues. In addition, it will provide better options for monitoring the status of power delivered to consumers. SMs provide a powerful way of enhancing power transaction process between the source and the sink in an SG. The functionalities of SMs in SGs vary depending on the application objective such as energy demand saving, feedback to consumers, dynamic pricing, and appliances control depending on demand curves, security enhancement, outage management, supply quality assessment [7,8], and demand response management schemes [9]. Hence, the SG is the ultimate solution to most of the challenges in current power grids. In fact, statistical studies show that the component failures in a power system can cause more than 80% of the electricity outages/cuts in a power distribution grid [10,11]. Hence, the SG will be able to adhere to such challenges, when it is properly planned. Taking the aforementioned remarks into consideration, this study aims to provide a comprehensive review on the role of SMs in SGs with a focus on the

PQ and PR monitoring applications by comparing ongoing attempts in literature while considering the different metrics and assessment standards. First, we present an overview of the AMI technology to provide an in-depth understanding of the general structure of an SG that consists of SMs, CTs, and RAs. Based on this comprehensive review, we also outline the open research issues in this field as possible future research directions.

The organization of this chapter is as follows. Section 1.2 reviews the related academic surveys presented in literature and outlines the contributions of this chapter. Section 1.3 discusses the key enabler technologies (SMs, CTs, RAs) for achieving a successful AMI for SGs. Sections 1.4 and 1.5 compare SM-related literature considering different metrics in the domains of PQ and PR, respectively. Sections 1.6 reveals the open research gaps for directing future researches, and the last section concludes the thoughts introduced in this chapter with some possible future work directions. In the following, Table 1.1 provides the definitions of used abbreviations in this chapter for more readability.

1.2   OVERVIEW OF RELATED SURVEYS

A number of attempts in literature have overviewed the usage of SMs and AMI technology in SGs. For instance, in Ref. [6], the authors review the challenges and advantages of integrating SMs. They further discuss the status of smart metering technology back then and believe that the main aim of SMs is to fight the basic problems which exist in power systems, rather than providing a luxurious operation scheme. Similarly, in Ref. [12], the authors review smart energy meters – namely electricity, heat, and gas meters – by shedding the light on various possible applications and benefits. Furthermore, the authors in Ref. [13] review the rising smart meter trends including AMI and CTs at MV and LV levels. This particular review sheds light on outage management – as a part of power reliability enhancement – in which an OMS is embedded into the communication structure. In another survey [7], Gouri R. Barai et al. discuss SM, AMI, and CTs as well as the benefits and challenges when it comes to SM integration. They briefly mention the usage of SMs for PQ and PR purposes. Another review in Ref. [14] discusses the elements of an SG and smart metering. In addition, the authors also discuss the status of SG development in some countries. In Ref. [15], the authors provide a comprehensive framework for the SG apparatus including SMs data processing, AMI, CTs, and brief discussion on RAs, and PQ and PR. Saket Nimbargi et al. review the AMI technology and the status of SM development in various countries [16]. They also consider AMI standards and cost estimates considering Indian protocols. In Ref. [17], the authors reviews the CTs used in smart meters as well as some network deployment schemes with a look at the Indian vision for communication architecture for smart metering. In Ref. [18], the authors review selected SMs' functionalities with an intense focus on data analysis aspects such as complexity, collection speed, and volume of data. They also survey SOM, SVM, and FL as data analysis techniques used in SMs. In 2018, two reviews [19,20] targeted different important aspects in SGs, one of which [19] illustrated a communication network structure designed for energy theft identification. In addition, they summarized the existing techniques used in literature for the same target, theft identification. On the other hand, in Ref. [20],

TABLE 1.1
Used Abbreviations

Abbreviation	Definition	Abbreviation	Definition	Abbreviation	Definition
ACO	Ant colony optimization	$CEMSMI_n$	Customer experiencing multiple sustained interruption and momentary interruption events index	HV	High voltage
ACRA	Artificial cobweb routing algorithm	CT	Communication Technology	ILWT	Integer lifting wavelet transform
AMI	Advanced metering infrastructure	CTAIDI	Customer total average interruption duration index	IoT	Internet of Things
AMR	Automatic meter reading	DNRPS	Dijkstra-based dynamic neighborhood routing path selection	LAN	Local area network
ANN	Artificial neural networks	DWT	Discrete wavelet transform	LV	Low voltage
ASAI	Average service availability index	EENS	Expected energy not supplied	MAIFI	Momentary average interruption frequency index
ASIDI	Average system interruption duration index	EQRP	Energy-efficient and QoS-aware routing protocol	$MAIFI_E$	Momentary average interruption event frequency index
ASIFI	Average system interruption frequency index	FET	Fault and error tolerance	MCU	Microcontroller
BMO	Bird mating optimization	FFT	Fast Fourier transform	MLRA	Maximum-likelihood routing algorithm
CAIDI	Customer average interruption duration index	FL	Fuzzy logic	MV	Medium voltage
CAIFI	Customer average interruption frequency index	FPGA	Field-programmable gate array	NAN	Neighborhood area network
CELID	Customer experiencing long interruption duration	GA	Genetic algorithms	NDN	Named data networking
$CEMI_n$	Customer experiencing multiple interruptions	GBR	Greedy backpressure routing	NFMCR	Neurofuzzy-based optimization multi-constrained routing

(Continued)

TABLE 1.1 (*Continued*)
Used Abbreviations

Abbreviation	Definition	Abbreviation	Definition	Abbreviation	Definition
OMS	Outage management system	RPA	Recursive pyramid algorithm	SVM	Support vector machine
OSPF	Open shortest path first	RTC	Real-time clock	THD	Total harmonic distortion
PDR	Packet delivery ratio	SAIDI	System average interruption duration index	TTRP	Transmission time for remaining path
PLR	Packet loss rate	SAIFI	System average interruption frequency index	WAN	Wide area network
PQ	Power reliability	SG	Smart Grid	WMT	Wavelet multiresolution
PR	Power reliability	SLPR	Straight-line path routing	WPT	Wavelet packet transform
PSO	Particle swarm optimization	SM	Smart meter	WSN	Wireless sensor network
QoS	Quality of service	SOM	Self-organizing map	WT	Wavelet transform
RA	Routing algorithm	ST	S-Transform		

TABLE 1.2
Summary of Related Surveys

References	Year	SM	AMI	CTs	RA	PQ	PR
[20]	2018	✓	✓	✓	✗	✗	✗
[19]	2018	✓	✓	✓	✗	✗	✗
[18]	2017	✓	✓	✓	✓	✗	✗
[17]	2017	✓	✓	✓	✗	✗	✗
[16]	2017	✓	✓	✓	✗	✗	✗
[15]	2017	✓	✓	✓	✓*	✓*	✓*
[14]	2017	✓	✓*	✗	✗	✗	✗
[7]	2016	✓	✓	✓	✗	✓*	✓*
[13]	2016	✓	✓	✓	✗	✗	✓
[12]	2015	✓	✗	✓	✗	✗	✗
[6]	2011	✓	✓	✓	✓*	✗	✗

* Briefly mentioned.

the author reviews the AMI technology and the communication structure for SGs in four different domains – namely, operation, transmission, distribution, and customer domains – with an in-depth focus on privacy and security issues.

However, none of these attempts have comprehensively targeted RA, PQ, and PR aspects in the context of the SG. These topics have been occasionally mentioned in different articles. However, no in-depth details have been provided. Table 1.2 compares the aforementioned references in relation to the contents of this work. It is clear that the RA, PQ, and PR aspects need further attention. Thus, our main intention is to provide a comprehensive review on the available literature targeting those areas. Our contributions in this work can be listed as follows:

- We aim at providing an in-depth understanding for the SM-based AMI technology in SG applications considering key enablers such as SMs, CTs, and RAs.
- We discuss significant design factors in wireless CTs as well as RAs for data processing in the AMI paradigm.
- We overview and categorize the existing solutions, communication technologies, and artificial intelligence techniques that target PQ and PR assessments using AMI.
- Finally, we suggest future and potential research directions in the SGs.

1.3 ADVANCED METERING INFRASTRUCTURE (AMI) TECHNOLOGY

The recent vast shift towards intermittent renewable energy generations, the continuous increase in power demand, and the environmental issues related to conventional energy sources are all considered as challenges in conventional power

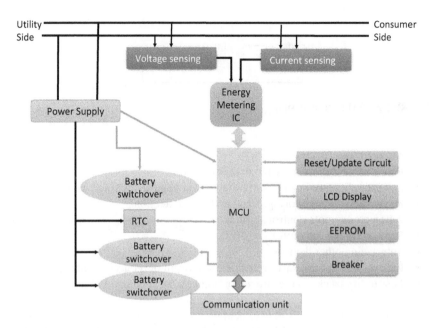

FIGURE 1.1 General structure of AMI technology in SGs.

systems. Further, in the conventional power system, it is difficult to have complete information on the power flow in many aspects such as power quality, reliability, and energy usage at different loads. With that being said, new technologies can be employed in a synergetic and integrated manner, modernize existing power systems, and cope with the challenges. AMI offers a sustainable solution in this regard which provides a two-way communication scheme between utilities and loads (consumers) as shown in Figure 1.1. Data including voltage, current readings, and demand curves will be collected from loads using SMs, and then, will be transferred using AMI to clouds and then to utilities in order to process the data and manage transmission and distribution processes. Then, a feedback is sent to consumers in order to monitor their consumption patterns and check the quality of received power.

1.3.1 SMART METER INTERNAL STRUCTURE

A key enabling device in the AMI is the SM, where it is installed on the consumer side for collecting real-time voltage and current data. Unlike conventional AMR where data collection is monthly, SMs provide the ability of daily data collection [16] via communication networks. Hence, SMs in SGs are beneficial not only for consumers but also for utilities and environment. The major features of SMs, but not limited to, are listed below [7,21]:

- Energy billing
- Electricity consumption reduction
- Consumption curves for both ends

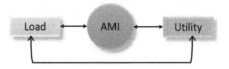

FIGURE 1.2   SM internal design.

- Net metering
- Power reliability monitoring: outage detection
- Power quality monitoring: harmonics and voltage disturbances classification
- Power security monitoring: fraud and thief detection
- Automated remote control abilities
- Remote appliance control
- Interfacing other devices
- Indirect greenhouse gases reduction as a result of reduced demand
- Less utility trucks in the streets for outage allocation and PQ tests.

This list implies that AMI is able to deal with most of conventional power systems' challenges relative to the AMR technology. It can be said that it is expected to have more complexities in the structure of SMs since it requires integration of high-tech components to provide good functionalities and features as illustrated in Figure 1.2 [22].

SMs mainly consist of an MCU, a power supply unit with a complimentary battery, voltage, and current sensors for active and reactive energy measurement in the energy metering IC, a RTC, and a communication facility, which are described in the following [23].

1.3.1.1   Microcontroller

Microcontroller unit (MCU) is the heart of an SM where most of the major data processing occurs. Therefore, all operations and functions in the SM are controlled by the MCU including the following:

- Communication with the energy metering IC
- Calculations based on the data received
- Display electrical parameters, tariff, and cost of electricity
- Smartcard reading
- Tamper detection
- Data management with EEPROM
- Communication with other communication devices
- Power management.

Nowadays, most of SMs are equipped with LCD interfaces that enable the consumers to not only learn their electricity tariffs and energy consumption patterns but also learn the quality of power delivered from utilities as well as the indication of a power outage when it occurs. The MCU also processes such functionalities.

1.3.1.2 Power Supply Unit

The SM circuit is supplied with power from the main AC lines through AC–DC converters and voltage regulators. A supplemental switchover battery is charged from the main AC lines in order to power the circuit when the connection between main AC and power supply unit is interrupted or a power outage occurs. Solar cells and rechargeable batteries can also be used to supply SM with power during the day [24].

1.3.1.3 Energy Measurement Unit

Based on the voltage and current readings sensed by the voltage and current sensors, energy measurement units perform signal conditioning, and computation of active, reactive, and apparent powers. Energy measurement units can operate as an embedded chip into the MCU or as a standard separated chip to provide the measurements as voltage or frequency pulses.

1.3.2 Machine Learning in AMI

Machine learning is another important component in AMI. Many sensors integrated in an SM are added to the smart grid for more efficient AMI implementation. The vast amount of information collected from the end users is valuable for researchers and the smart grid operators as well. And hence, an advanced analytics on the smart grid is needed, where a combination of machine learning algorithms and data mining techniques is applied. By exploiting the emerging smart grid collected data, we can develop data-driven solutions for the most pressing issues, such as electricity demands prediction per region in a smart city, residential photovoltaic detection, electrical vehicle charging demand determination, and the time-variant load management problem.

Machine learning deals with the gather grid information in order to provide the SG the ability to learn from its history like humans. It provides information about the properties of the collected data, allowing it to make predictions about other data that may occur in the future. Generally speaking, there exist three main categories of learning in the SG, which are supervised, unsupervised, and semi-supervised learning algorithms.

1.3.2.1 Supervised Learning Algorithms

These algorithms use training data to generate a function that maps the inputs to desired outputs (also called "labels"). For example, in a classification problem, the system looks at sample data and uses it to derive a function that maps input data into different classes. Artificial neural networks, radial basis function networks, and decision trees are other forms of supervised learning.

1.3.2.2 Unsupervised Learning Algorithms

This set of algorithms work without previously labeled data. The main purpose of these algorithms is to find the common patterns in previously unseen data. Clustering is the most popular form of unsupervised learning. Hidden Markov models and self-organizing maps are other forms of unsupervised learning.

1.3.2.3 Semi-Supervised Learning Algorithms

As the name indicates, these algorithms combine labeled and unlabeled data to generate an appropriate mapping function or classifier. Several studies have proven that using a combination of supervised and unsupervised techniques instead of a single type can lead to much better results.

1.3.3 WIRELESS COMMUNICATION IN AMI

Data communication in AMI is an essential part where data are instantly collected and transferred to the utility to process it, and then, utilities send a feedback to consumers accordingly. An SG covers a large geographical area, and hence, the communication structure is clustered into regions in order to assure QoS in data transfer. Communication areas can be divided into three main regions as given in Ref. [25]. The first is LAN which describes the communication scheme between consumers and SMs. The second is the NAN region which basically represents communication medium that contains flowgates to perform specific processes (such as data aggregation and encoding) on the data coming from SMs before it is transmitted to the cloud. The third region is WAN which is responsible to communicate data between the cloud and destination (utility). Figure 1.3 illustrates the communication structure in SGs.

However, there are specific areas that need further considerations in AMI communication. In AMI infrastructure, big data transmission, data security, network scalability, and cost effectiveness are among the essential areas that need more attention [21]. Thus, there is a need for international standards and regulations to put a

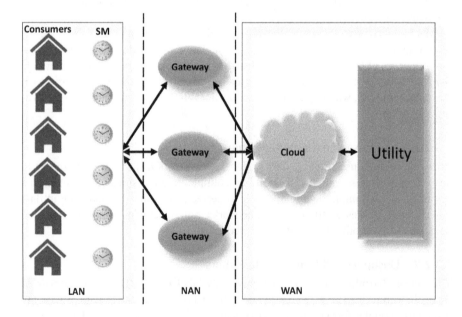

FIGURE 1.3 A review of the categorized communication technologies in SGs.

framework on communication aspects in AMI. In this regard, there are various standards developed by international institutions such as IEEE 802.15.4, IEC 61970, and ISO 1802 [26] in order to insure reliable, secure, and efficient power delivery to the consumers. By analyzing the literature, there have been a variety of communication protocols developed [27,28] according to international standards that can be used in LAN, NAN, and WAN regions of the AMI in SGs. The most common used wireless communication protocols in AMI are summarized in the following.

1.3.3.1 Local Area Network (LAN)

1.3.3.1.1 Zigbee

Zigbee is a communication technology developed according to the IEEE 802.15.4 standard that transmits data at a rate of approximately 250 Kbps on 2.4 GHz frequency [29]. The technology targets applications that require a short range of communication, 0–100 m. Despite the short range of coverage, Zigbee proves itself in its low-power consumption characteristic as well as system scalability at the cost of data transmission rate. Hence, this technology can show a good performance in the LAN region of an SG.

1.3.3.1.2 Wi-Fi

Wireless fidelity (Wi-Fi) is another communication technology that is largely used nowadays in homes and business areas. It is designed according to the IEEE 802.11b/g/n standard with a capability of data transmission at frequencies 2.4 and 5 GHz in the domain of 0–250 m at a rate of 54 Mbps [30]. The domain of Wi-Fi is larger compared to Zigbee. It also provides a relatively higher rate of data transmission at the cost of scalability. Wi-Fi technology is mainly integrated into the SG system as an in-home networking scheme inside the aforementioned LAN area. This is due to its security advantage since it has a robust authentication/access procedure.

1.3.3.1.3 Bluetooth

Bluetooth is another technology that can be used for in-home networking in the SG system since it has a limited low energy communication range of 0–100 m. This technology is developed according to the IEEE 802.15.1 standard to transfer data at a rate of 721 Kbps [31] and a frequency of 2.4 GHz. Bluetooth is distinguished by its low-power consumption, and thus, it has a limited capability of vast data transfer. The widespread use of the technology among users especially in smartphones gives it another advantage to be integrated in SG LANs. In addition, it's ideal for connecting the LAN of the in-home appliances with the SM.

1.3.3.1.4 Z-Wave

The Z-wave technology is a radio frequency-based communication mainly designed for in-home appliances remote control which relies on an unlicensed 900 MHz frequency which transmits data at a rate reaching 40 kbps for a range of 0–30 m [32]. The operating frequency of this technology reduces the risk of being disrupted by the previously mentioned technologies since they operate on 2.4 GHz frequency, which enhances the reliability of this technology. Since the data transmission speed is a concern in the SG, this Z-wave technology is tested and proved to have low delay rates (in milliseconds) even with congestion that can reach 1000 simultaneous client requests in the LAN/HAN network [33].

1.3.3.1.5   NFC

Near-field communication (NFC) is a set of short-range protocols working at the range of about 10 cm, and its targets can be simple devices such as stickers, cards, or unpowered tags attached to the SM in an AMI [34]. In addition, NFC allows peer-to-peer communication as well in which both devices should be powered. It can be utilized in realizing an electronic verification of the utility permits which in turn allows providing better services to citizens. It provides low-power and low-cost wireless connectivity within short ranges of up to 20 m which makes it suitable for use in WSNs, M2M, and IoT-enabled SG.

1.3.3.2   Neighborhood Area Network (NAN) and Wide Area Networks (WAN)

1.3.3.2.1   NB-IoT

The NB-IoT technology plays an important role in the AMI of an SG. The technology consists of 2G, 3G, and 4G cellular modes that provide the capability of big data transmission with rates ranging from 14.4 kbps (for 2G) to 100 Mbps (for 4G) at a licensed frequency band (824 and 1900 MHz) [35]. The technology targets a large area because of its long domain that ranges between 10 and 100 km. With that being said, NB-IoT technology consumes high power for the transmission process.

1.3.3.2.2   Sigfox

Sigfox is a developing machine-to-machine WAN communication solution that operates on a frequency band of 868 MHz that has the ability to cover 30–50 km in rural areas and 3–10 km in urban areas to deliver the data at a rate of 100 bps [36]. An advantage with this technology is noticeable in its low power consumption for data transmission. The limitation of Wi-Fi technology to be applied in NAN or WAN is the short range of coverage. On the contrary, NB-IoT provides a solution for big data transmission in WANs at the cost of power consumption. Hence, Sigfox seems to provide a middle-way solution considering the range of coverage and power consumption in comparison with NB-IoT.

1.3.3.2.3   LoRaWAN

LoRaWAN is a recent non-profitable organization that is mainly established to be integrated in Internet of Things (IoT) WAN applications. The technology operates at 900 MHz frequency to transmit data at a rate of 50 kbps for a distance ranging between 10 and 15 km in rural areas and 2 and 5 km in urban areas [37]. Relative to Sigfox technology considering the trade-off between coverage and power consumption, LoRaWAN offers a better data transmission rate with reduced power consumption. Table 1.3 illustrates a summary of communication technologies used in LAN, NAN and WAN areas of an SG.

TABLE 1.3

Categorization of the Communication Technologies in an AMI

Application in SGs	Technology	Data rate	Coverage	Frequency	Standard
LAN	Zigbee	250 kbps	0–100 m	2.4 GHz	IEEE 802.15.4
	Wi-Fi	54 Mbps	0–250 m	2.4 and 5 GHz	IEEE 802.11b/g/n
	Bluetooth	721 kbps	0–100 m	2.4 GHz	IEEE 802.15.1
	Z-wave	40 kbps	0–30 m	900 MHz	ITU-T G. 9959
NAN & WAN	NB-IoT	14.4 kbps (2G) 100 Mbps (4G)	10–100 km	824 MHz and 1900 MHz	GSM/GPRS/EDGE, (2G), UMTS/HSPA (3G), LTE (4G)
	Wi-SUN	300 kbps	500 m–5 km	900 GHz	IEEE 802.15.4g
	Sigfox	100 bps	30–50 km (Rural) 3–10 km (Urban)	868 MHz	Sigfox
	LoRaWAN	50 kbps	10–15 km (Rural) 2–5 km (Urban)	900 MHz	LoRaWAN
	LoRa modulation	0.3–37.5 kbps	10–15 km (Rural) 3–5 km (Urban)	900 MHz	LoRa modulation

1.3.3.2.4 Wi-SUN

This communication technology is developed in line with the IEEE 802.15.4 g standard in which it operates at a low frequency of 900 MHz to transmit the collected data at a rate of 300 Kbps for an area domain of approximately 500 m to 5 km [38]. Wi-SUN provides a higher data transmission rate in comparison with LoRaWAN and NB-IoT 2G technologies. The usefulness of this technology is seen in low-latency communications that make it a good choice for SG NAN/WAN applications where simultaneous data processing is required.

1.3.3.2.5 LoRa

LoRa is a physical layer in low-power wide area network (LPWAN) solution designed by Semtech Corporation which manufactures the chipsets as well [39]. LoRa technology contains two main components. One of them is the LoRaWAN network protocol which has been mentioned earlier and is optimized for energy-limited end devices [39]. The other one is *LoRa modulation* which is based on chirp spread spectrum technique that utilizes wideband linear frequency modulated pulses in which the frequency increases or decreases according to the encoded information [39]. In addition, *LoRa modulation* consists of a variable cyclic error correction scheme that enhances the performance of the system by adding some redundancy [39]. The coverage range of LoRa is 10–15 km in rural areas and 3–5 km in urban areas. Moreover, in terms of data rates, LoRa has a data rate between 0.3 and 37.5 kbps [39].

1.3.4 ROUTING ALGORITHMS

One of the essential SG characteristics that differ from conventional power grids is communication network. The new setup of communication networks in SGs should be distinguished by their ability to support time-sensitive and data-intensive management tasks [40]. For instance, the closer the SMs to the gateways, the more the information they will transfer which leads to more data concentration which may affect the transmission reliability. Therefore, choosing the appropriate routing algorithm is an essential area to be covered in SG communication network setups [41]. Data routing is moving one data unit through one possible communication path or more from the source to the destination node [42]. Having multiple paths to transmit the data unit from the source to the destination increases the network robustness relative to one single path in which there is a higher possibility for the path to fail. Therefore, a robust routing scheme implies adding more data transmission complexity. Since there may be multiple paths for data transmission in SG communication networks, a routing algorithm decides on which path the data will follow considering different metrics. It can be said that reliability, cost, computational power, delay, and data throughput are some of the key objectives when developing a routing algorithm [43] for SG applications. The following discusses major RA design objectives.

1.3.4.1 Delay

In SG communication networks, if the transmitted information from the source to the destination is received late, undesired events might happen even though the data control devices take a correct action [44]. Therefore, minimizing delays in the communication network is essential to enhance the quality of data delivery. Route selection is the essence of limiting the delay in SG communication networks. The authors in Ref. [44] estimate the delay and propose the minimum end-to-end delay multicast tree routing algorithm to select the path with least delay in NAN region of SGs. Another routing algorithm is developed in Ref. [45] to ensure an efficient communication path selection for SG applications in the NAN communication region. The DNRPS algorithm shows its superiority in the reduced communication delay by a highest gain percentage reduction of 19% over other selected algorithms, and Dijkstra is one of them. Delay is also studied in Ref. [46] where the authors use the ACRA to enhance the QoS of the network by focusing on the delay and data throughput for applications in low-voltage power distribution networks, in this context, LAN region. Moreover, in Ref. [47], the author propose a routing algorithm for QoS enhancement considering delay, memory utilization, packet delivery ratio, and throughput using EQRP algorithm that is inspired by BMO. Furthermore, the authors in Ref. [48] attempt to study the delay for SG applications using greedy QoS routing algorithm, whereas in Ref. [49], the SLPR algorithm is used. Delay in some works [50,51] is referred to as latency. In Ref. [50], the author uses the layered cooperative processing algorithm to enhance the QoS based on latency and reliability for SG application, whereas in Ref. [51], the latency and reliability are objectives enhanced using an ACO algorithm for the application in NAN region. Moreover, in Ref. [49], the authors use the GBR algorithm to study latency as well as throughput. Following the analysis on this literature, various algorithms are used for different application regions in the SG (LAN, NAN, or WAN), and some studies target the SG in general.

1.3.4.2 Security

Smart meters autonomously collect massive amounts of data and transfer it to the utility company, consumer, and service providers. This data includes private consumer information that might be used to infer consumer's activities, devices being used, and times when the home is vacant. Moreover, the smart grid has several intelligent devices that are involved in managing both the electricity supply and network demand. These intelligent devices may act as attack entry points into the network. Unlike the traditional power system, smart grid network includes many components, and most of them are out of the utility's premises. This fact increases the number of insecure physical locations and makes them vulnerable to physical access. Unfortunately, there are outdated equipment which are still in service and coexist with the grid. These equipment might act as weak security points and might very well be incompatible with the current power system devices. In addition, having many stakeholders might give raise to a very dangerous kind of attack called the "insider attacks." Accordingly, a careful attention shall be paid to security issues in the smart grid before being realized in practice.

1.3.4.3 Coverage

Wireless connectivity is a cost-effective solution to modernize the electrical grid turning it into a smart grid system. Deploying a wireless communication system is easier for consumers than installing wired communication meters. Wireless allows two-way digital communications by adding computer intelligence and data communications to the electricity distribution networks from non-renewable (coal and nuclear) and renewable energy (solar and wind) to smart appliances to plug-in cars. Key components that enable smart grid to provide two-way wireless controls and communications are smart (wireless) meters, backhaul network, utility pole radios, central utility center, and the remote sub-stations. Technical specifications and bandwidth capacity per component in the above communication system specify the coverage region of the smart grid. And hence, the more powerful components we have, the more coverage we can gain. In general, the key characteristics that influence the overall coverage range of a smart grid can be listed and quantified using the Friis transmission formula as follows: (1) transmission RF power, (2) antenna gain on transmitter and receiver, (3) frequency band which can vary from 700 MHz to 5.85 GHz, (4) receiver power sensitivity, and (5) path loss effect.

1.3.4.4 Scalability

Scalability is a central issue in the development and deployment of a smart grid system. The scalability footprint of a smart grid solution can be defined by two core criteria: load and complexity. Load is the more poignant factor, and in order to isolate it, complexity has to be minimized if possible. Load scalability encompasses memory, communication, and CPU loads. It is suggested to implement a smart grid with a hierarchical-based structure in order to ease/relax the scalability issues where better communication alternatives can be utilized unlike the meshed structure for example. Because scalability is a central player for the grid performance, it is important to examine how the different grid architectures resist its influence as the system consumers increase. To prevent distorting the structure of these architectures, the increased grid scale can be achieved through the injection of additional smart meters. Smart meters' quantities shall be divided equally among the demand clusters in the served region for better load balance and fair solutions.

1.3.4.5   Firmware Updates

Security in IoT can easily get broken, and the connected things including SMs can get hacked, especially if there are lots of them. For that reason, it is of utmost importance that these devices must be able to upgrade over the air (OTA). Moreover, OTA updates can significantly increase the SG scalability and deployment reliability. For example, when something goes wrong, a new firmware image (the embedded software which controls the thing) can be sent over a wireless connection to replace the broken/nonfunctioning firmware. Doing it wirelessly removes the need for a man power to be sent out, which can dramatically slow down the service, especially if we are talking about upgrading millions of these SMs.

Meanwhile, upgrading an SM would be easy, if it has a high-speed wireless connection. However, SMs typically send a few hundred bytes of data to the cloud and rarely need any data to be sent back. And hence, current SG networks are optimized for data being sent from meters, not to them, which forms another challenge in upgrading any SM as they are not designed to send firmware update files to millions of meters at the same time.

1.3.4.6   Lifetime

SMs in the communication network of an SG are scalable, flexible, and intelligent nodes. However, these sensor nodes are subject to failure due to their limited energy capacities, which shortens the overall network lifetime. For this, the authors in Ref. [52] propose an energy routing-based WSN scheme to lengthen the network lifetime considering the sensor node energy limits for applications in power distribution networks, whereas in Ref. [53], the authors show the advantage of using MLRA over random method for extending the lifetime of SG wireless network.

1.3.4.7   Cost

Routing cost is another important objective that is essential when designing a routing algorithm. It is proposed in Ref. [54] that relying on OSPF technique can help reduce the cost of route selection for applications in LANs of SGs. On the contrary, in Ref. [55], the authors rely on FL, namely NFMCR algorithm, for cost and error reduction in route selection for application in SGs. Cost alongside fault tolerance is also studied in Ref. [56] where the authors employ the Dijkstra shortest path routing algorithm considering all SG communication regions, LAN, NAN, and WAN. In another study [57], the authors also use the Dijkstra's shortest path routing algorithm to find the least routing cost among the network nodes putting the security of the grid into consideration.

1.3.4.8   Reliability

Reliability in communication networks has something to deal with failures when transmitting information from the source to the destination [58]. From the hardware point of view, reliability can be enhanced through the inclusion of more links and

TABLE 1.4

Categorization of Routing Algorithms According to Objectives in SG AMI

References	Objective								
				Reliability					
	Delay	Lifetime	Cost	PDR	PLR	FET	Scalability	Coverage	Security
[44,45,48,49]	✓							✓	
[46,49]	✓						✓		✓
[47]	✓			✓			✓	✓	
[50]	✓	✓		✓					✓
[51]	✓				✓			✓	
[52,53]		✓					✓		✓
[54,57]			✓					✓	
[55,56]			✓			✓			
[60]						✓		✓	✓
[62]				✓	✓				
[63]		✓					✓		✓

components to overtake the work of faulty components in the failure situation [59] which increases the complexity and installation costs. Hence, a reliable communication network is fault/failure and error tolerant, FET. Some studies attempt to improve network reliability considering error reduction [55], fault tolerance [56], or failure probability [60]. Some other studies also examine the PDR as an indicator of reliable communication. PDR is the ratio between the amount of data packets sent by a source and data packets received at the destination [61]. In an SG domain, PDR is improved in a study [47] using EQRP as mentioned previously. Moreover, the reliability of the link between two nodes in reference [50] is assessed using PDR, and reliability routing decision algorithm is used to select the reliable routing path. Similar to PRD, in Ref. [51], the authors attempt to reduce PLR using an improved ACO for applications in SG NAN regions. Furthermore, both PDR and PLR are considered in Ref. [62] for applications in NAN using hybrid metric algorithm. Based on literature, it can be concluded that reliability enhancement is essential and should be studied in SG LAN networks since data congestion is higher on SMs around gateways. Table 1.4 summarizes related studies based on their design objectives.

1.4 SMART METERS AND POWER QUALITY

Power quality simply means delivering a smooth and a steady voltage waveform to consumers. PQ issues are majorly found in variant voltage waveforms and supply frequency. These waveform disturbances accompanied with current and voltage distortions can be caused by changeable load patterns [4]. Hence, it is essential for utilities to detect and classify those disturbances, as it will enhance the power grid performance. Using SMs in assessing PQ is an important switching point as it implies the application of SGs measures. In the following subsections, we provide an overview of the literature works that target the employment of SMs into PQ assessment.

1.4.1 ASSESSMENT PARAMETERS

The global importance of PQ assessment in power distribution networks [64–66] brought a great attention to standardizing PQ guidelines in order to promote its assessment accuracy. Many organizations successfully introduced international standards for PQ assessment. Institute of Electrical and Electronics Engineers (IEEE) [67] and the International Electro-technical Commission (IEC) [68] are among the reputable organizations that significantly contributed in developing PQ standards. For instance, IEEE 519-1992 is an IEEE-recommended practice for harmonics control in electrical power systems intended to provide steady-state operation limits in order to minimize harmonics and transients [69]. This practice is widely adopted by North American power utilities [70]. Subsequently, IEEE 1159-1995 emerged aiming to build a guideline for acceptable methods of monitoring PQ in power distribution networks [71]. In addition, it classifies the typical characteristics of electromagnetic phenomena parameters that mainly cause PQ. Two standards (IEC 61000-4-7 and IEC 61000-4-15) developed by the IEC already exist [72] in which they included power quality parameters along with their calculations and interpretation methods. Further developments continued to introduce the IEC 61000-4-30 standard [73]. Furthermore, EN50160 is a European power quality standard that is adopted by many European countries [74]. It is essential to consider standardized parameters while assessing PQ. The common parameters introduced by international standards are used in this chapter for the comparison to draw a clearer picture of what mostly causes poor power quality.

Technological advancements are essential for PQ assessment [75] as it needs capable devices and instruments and algorithms to achieve accurate monitoring that meets the required standards and regulations [75,76]. AMI technology plays an important role in this regard. The authors in Ref. [77] attempt to develop an SM network for electrical installation monitor for PQ assessment parameters including power factor, THD, and voltage dips, swells and interruptions. They consider an in situ big data processing capability relying on an FPGA embedded in the SM. But in [78], they introduce a PQ assessment SM putting cost of measurement device into consideration. They use voltage transient detection, current drop patterns, and arc-fault detection as their evaluation parameters. In Ref. [79], voltage distortion and imbalance were used to evaluate PQ and provide an advanced warning of PQ problems as well as free data processing capability. Furthermore, a smart monitoring system is proposed in Ref. [80] that relies on voltage deficiency indications to reduce customer complaints and operational costs. In Ref. [81], the authors consider reduction of computational effort for PQ disturbances classification such as sags, swells, flickers, harmonic distortion (HD), voltage interruptions, and oscillatory transients. That is to say, studies can be categorized according to the parameter(s) used in their assessment criteria as shown in Table 1.5.

TABLE 1.5

Studies Categorization According to Used Assessment Parameters

References	Transients	Sags	Swells	Under/over Voltage	Voltage Unbalance	Voltage Interruptions	Dips	Power Frequency	Flicker	Reactive Power	Harmonics
[77]	✓					✓	✓				✓
[78]			✓					✓			✓
[79]					✓						✓
[80]				✓							
[81]	✓	✓	✓			✓			✓		✓
[82]										✓	
[83]											✓

1.4.2   TECHNIQUES EMBEDDED IN SM FOR PQ ANALYSIS

Classification and detection of PQ disturbances have been an important topic that many researchers continuously attempt to solve [81]. Various methods have been introduced in the literature to efficiently provide the status of disturbances in a power system. Regardless of using SMs in power quality assessment, the commonly used techniques are based on either signal processing techniques (FFT, WT, ST, etc.), artificial intelligence approaches (ANN, FL, and SVM), or heuristic optimization approaches (PSO, GA, etc.) [81,84,85]. However, these techniques and studies presented in the literature focus on PQ assessment in the context of conventional power grid. In other words, with the emergency of SG's definitions and measures accompanied with the numerous deployments of SMs, PQ analysis techniques should shift towards smart ways of implementation. A key factor of achieving the smartness of PQ analysis can be introducing in SMs-embedded analysis techniques. Undoubtedly, this will enable having SMs with multifunctionalities alongside PQ assessment such as load profile monitoring, energy billing, outage detection, or even remote automated switch control in SGs. In this context and based on the limited literature related, two techniques are revealed and compared next.

1.4.2.1   Wavelet Transform (WT)

WT is a mathematical model that plays an important role in signal analysis for the purpose of PQ assessment as it provides the ability to analyze waveforms' characteristics in time–frequency domain [85]. The WT is useful when voltage transients as well as short-duration voltage variations (sags, swells, interruptions) are considered in the PQ studies [86]. In the light of techniques embedded in SMs for PQ studies, it is worth mentioning selected studies. For instance, in Ref. [82], the authors use the DWT to measure reactive energy with the presence of time-variant PQ disturbances like voltage swells and harmonics for the objective of maintaining a less computational effort relative to WPT, whereas in Ref. [78], a WMT approach is used to detect voltage transients and current drops as well as a THD measurement Goertzel algorithm considering low smart meter cost. A developed technique relying on employing the RPA in applying DWT was introduced in Ref. [87] where the primary aim is showing the computational efficiency of the proposed model over the normal DWT. Following an analysis presented in Ref. [88], a novel ILWT technique is used with SMs to achieve real-time compression and transmission of signals for the purpose of analyzing harmonics and PQ short-duration disturbances such as sags, swells, voltage transients and interruptions, and flickers.

Following the analysis of literature, it can be said that the main advantage of using WT is observed in its ability to provide a good analysis resolution in the time–frequency domain, and hence, short-duration voltage variations can be classified as mentioned previously. However, WT can induce computational burden on MCUs in SMs (the DWT is more efficient in terms of computation effort as shown in Table 1.6) especially when better analysis performance is desired, which is a major disadvantage of this technique.

TABLE 1.6

Techniques Embedded in SMs for Different Applications and Objectives

Technique		Application(s)	Objective	References
WT	DWT	Reactive power	Computational effort	[82]
	RPA -DWT	Reactive power	Computational effort	[78]
	WMT	Voltage transients	Low device cost	[87]
	ILWT	Harmonics and disturbances	-	[88]
FFT		Harmonics	Computational effort	[80]
		Harmonics	Fast performance	[89]

1.4.2.2 Fast Fourier Transform (FFT)

FFT is another widely used signal analysis technique which mainly converts signals from time to frequency domain [90]. The usefulness of this transform is observed in periodical signals, i.e., identifying their phases and amplitudes [91] and hence determining noticeable harmonic events. As in Ref. [32], it is used to achieve a low computational burden on the hardware while evaluating the HD. In this study, the model is only applied on harmonics extraction, and two other artificial intelligent approaches (ANN and decision trees) are used to extract short-term PQ disturbances including sags, swells, and oscillatory transients. The simplicity of implementation in FFT makes it a noticeably fast processing technique relative to other complex techniques. In Ref. [89], FFT is preferred for its fast performance and accuracy to detect and estimate THD when embedded in an SM.

One advantage of using FFT in PQ assessment alongside with SMs is that it provides fast performance and accuracy for harmonics evaluation. Another advantage is that FFT is observed to be suitable to be embedded in SMs for PQ assessment compared to other techniques [81,89]. Nevertheless, FFT shows weak performance in terms of short-term variation detections and time–frequency domain resolution relative to WT. Table 1.6 shows the usages of both techniques embedded in SMs for different applications and objectives.

1.5 SMART METERS AND POWER RELIABILITY

Power reliability simply is related to the total electric interruptions in a power system that has to do with the full loss of voltage waveform unlike power quality which covers voltage sags, swells, and harmonics [92]. A highly reliable power system means that power is to the consumers all the time without any interruptions. However, power systems are not ideal and various factors can affect power system reliability, which means economic loss to both utilities and consumers. It is estimated that 80% of power reliability issues occur in the power distribution network [93]. Hence, in conventional power systems, the details of interruptions and outages are hidden. With technological advancements, AMI technology and data analysis techniques are proved to enhance the exposure of power system issues and measure the severity of interruptions [94] which enhances the movement towards the concept of an SG. In the light of SMs, there have been many ways and techniques used in literature to evaluate the reliability of the delivered power from utilities to consumers. This is achieved through determining defined reliability indices, whereas some other techniques in literature attempt to detect power outage interruptions in a power system. Various standards construct a framework for assessing power reliability in a power distribution network. The IEEE1366-2012 standard is among the widely used standards that presents indices can be used in assessing power reliability [95] which presents sustained and momentary interruption indices as well as load-based indices. Frequency, duration, and the extent of the interruption are essential parameters used in characterizing reliability of the power system [96]. Indices presented by IEEE1366-2012 are briefly categorized in Figure 1.4. Sensors are deployed with sufficient amounts in HV and MV networks, which enables the accurate monitor of power reliability events. With the introduction of AMI in LV networks, namely, SMs, there is a great opportunity of enhancing the monitoring capabilities on the LV side. That is to say, reliability indices can be accurately calculated with the help of SMs at the LV sides. In Ref. [94], the authors present a method of calculating temporospatial disaggregated reliability SAIDI relying on SM data. In this context, SMs also provide the ability of interruption time reduction relative to conventional meters. Replacement of conventional meters with SMs in a power distribution utility in Brazil is discussed in Ref. [97] considering SAIDI and EENS. The study shows a noticeable annual reduction of both SAIDI and EENS for the period between 2011 and 2015 where they achieved a 16.54% reduction in 2015. The study [96] implemented in Helsinki, Finland, shows that the help of SMs enabled utilities to achieve a percentage reduction of 50% in SAID index. In contrast, AMI technology has more focus in Ref. [98] which develops a Zigbee-based automated reliability system that is able to calculate

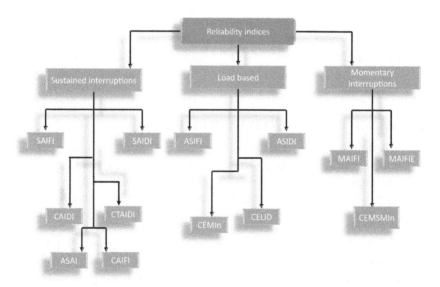

FIGURE 1.4   Summary of IEEE1366-2012 reliability indices.

TABLE 1.7

Summary of SM-Based PR Indices Calculation

References	Index(s)	Focus on SMs
[94]	SAIDI	Moderate
[96]	SAIDI	Moderate
[97]	SAIDI and EENS	High
[98]	SI and LB	High
[99]	PRI	High
[100]	ASIDI and ENS	Low

and display reliability indices including sustained and load-based indices. Following the focus on employing SMs in calculating reliability indices, the authors in Ref. [99] propose an evaluation model for power reliability index (PRI) which uses the PQ data recorded by the SMs. The usefulness of this study is observed in employing both PQ and reliability considerations to enhance the overall grid system operation with an intense focus on SMs, whereas in Ref. [100], the authors briefly consider the usage of SMs in this regard but attempt to calculate reliability indices such as ASIDI and ENS as an additional adjective. Following analysis of the limited literature in the domain of using SMs for PR indices calculations, applications vary in the indices calculated as well as the focus on the usage of SMs. Hence, it is essential to develop further attention on PR assessment using SMs considering either calculating PR indices or employing WSN schemes for outage detections. Table 1.7 summarizes the studies that rely on SM data for PR indices calculation.

1.6 OPEN RESEARCH ISSUES

According to analysis of previous literature in this chapter, we suggest further investigations on the following issues:

1. Big data management schemes in order to reduce data congestion in LAN.
2. Multifunctional SM developments. Additional functions can include PQ and PR assessments together with the typical SM tasks.
3. Optimal reconfiguration of communication infrastructure so that it accommodates PQ and PR assessments schemes through either calculating PQ/PR indices or employing WSN for event detection.
4. Simultaneous and optimization algorithms for network reconfigurations.
5. Development of ML algorithms embedded in SMs for real-system health monitoring, which can include various PQ and/or PR indices.
6. Development of SM-embedded techniques to detect short-time supply/demand variations.
7. Development of PQ and PR monitoring/assessment systems according to recognized world standard.
8. Development of efficient and accurate SM-embedded algorithms and techniques.

9. Development of synergetic integrated models that include other AMI applications alongside PQ and/or PQ monitoring.
10. The number of customers in SGs is huge and exponentially increasing. Hence, scalability is an essential design parameter as to enhance the communication network reliability.
11. The massiveness of the smart grid and the increased communication capabilities make it more prone to cyber-attacks. Hence, further security-based solutions are required.
12. Practical issues with SGs like device configurations and software updates are still in need for more attention to be better realized. Over-the-air updates, for example, can offer various benefits such as the product behavior monitoring, and increased SG scalability and deployment reliability. However, due to these issues, it is still unreliable and prone to several failures.
13. The expected future transition from conventional power grids to SGs will need topology planning in the LV side of the grid considering the range of coverage. For example, tree/cluster-based sensory networks can be applied for better coverage.
14. CT categorization enhancement based on their respective advantages and disadvantages in relation to the topology plan is also required.
15. Enhancement of CTs data processing in terms of time sensitivity and data intensity is significantly needed.
16. RA enhancement in relation to the future SG communication technology is required.

1.7 CONCLUSION

The importance of PQ and PR in power grids has brought an increased attention to developing new ways and techniques for their assessment. IoT is believed to provide a long-term solution to help solving such problems as well as providing the essential building blocks for enforcing the measures of SGs by using AMI and SMs technologies. For this, we reviewed the effectiveness of employing these technologies onto conventional power grids for PQ and PR assessments. The structures of AMI and SM technologies including wireless communication technologies as well as data routing algorithms are thoroughly discussed, and open research areas are suggested accordingly.

REFERENCES

1. P. Kukuča and I. Chrapčiak, "From Smart Metering to Smart Grid," *Meas. Sci. Rev.*, vol. 16, no. 3, pp. 142–148, 2016.
2. N. U. Hassan, W. Tushar, C. Yuen, S. G. Kerk, and S. W. Oh, "Guaranteeing QoS Using Unlicensed TV White Spaces for Smart Grid Applications," *IEEE Wirel. Commun.*, vol. 24, no. 2, pp. 18–25, Apr. 2017.
3. R. Morello, S. C. Mukhopadhyay, Z. Liu, D. Slomovitz, and S. R. Samantaray, "Advances on Sensing Technologies for Smart Cities and Power Grids: A Review," *IEEE Sens. J.*, vol. 17, no. 23, pp. 7596–7610, Dec. 2017.

4. F. Al-Turjman, "Mobile Couriers' Selection for the Smart-grid in Smart cities' Pervasive Sensing," *Elsevier Future Generation Computer Systems*, vol. 82, no. 1, pp. 327–341, 2018.
5. S. Darby, "Smart Metering: What Potential for Householder Engagement?" *Build. Res. Inf.*, vol. 38, no. 5, pp. 442–457, 2010.
6. S. S. S. R. Depuru, L. Wang, and V. Devabhaktuni, "Smart Meters for Power Grid: Challenges, Issues, Advantages and Status," *Renew. Sustain. Energy Rev.*, vol. 15, no. 6, pp. 2736–2742, 2011.
7. G. R. Barai, S. Krishnan, and B. Venkatesh, "Smart metering and functionalities of smart meters in smart grid - a review," in *2015 IEEE Electrical Power and Energy Conference (EPEC)*, 2015, pp. 138–145.
8. W. Tushar et al., "Smart Grid Testbed for Demand Focused Energy Management in End User Environments," *IEEE Wirel. Commun.*, vol. 23, no. 6, pp. 70–80, Dec. 2016.
9. W.-T. Li et al., "Demand Response Management for Residential Smart Grid: From Theory to Practice," *IEEE Access*, vol. 3, pp. 2431–2440, 2015.
10. Y. Jiang, C.-C. Liu, M. Diedesch, E. Lee, and A. K. Srivastava, "Outage Management of Distribution Systems Incorporating Information from Smart Meters," *IEEE Trans. Power Syst.*, vol. 31, no. 5, pp. 4144–4154, Sep. 2016.
11. A. A. Chowdhury and D. O. Koval, "Reliability Principles," in *Power Distribution System Reliability*. Hoboken, NJ: John Wiley & Sons, Inc., pp. 45–77.
12. Q. Sun et al., "A Comprehensive Review of Smart Energy Meters in Intelligent Energy Networks," *IEEE Internet Things J.*, vol. 3, no. 4, pp. 464–479, Aug. 2016.
13. E. Kabalci, "Emerging smart metering trends and integration at MV-LV level," in *2016 International Smart Grid Workshop and Certificate Program (ISGWCP)*, 2016, pp. 1–9.
14. P. Bansal and A. Singh, "Smart metering in smart grid framework: A review," in *2016 Fourth International Conference on Parallel, Distributed and Grid Computing (PDGC)*, 2016, pp. 174–176.
15. H. Daki, A. El Hannani, A. Aqqal, A. Haidine, and A. Dahbi, "Big Data management in Smart Grid: Concepts, Requirements and Implementation," *J. Big Data*, vol. 4, no. 1, p. 13, Dec. 2017.
16. S. Nimbargi, S. Mhaisne, S. Nangare, and M. Sinha, "Review on AMI technology for smart meter," in *2016 IEEE International Conference on Advances in Electronics, Communication and Computer Technology (ICAECCT)*, 2016, pp. 21–27.
17. R. Pillai and H. Thukral, "Next generation smart metering: IP metering," *CIRED - Open Access Proc. J.*, vol. 2017, no. 1, pp. 2827–2829, Oct. 2017.
18. S. Pawar and B. F. Momin, "Smart electricity meter data analytics: A brief review," in *2017 IEEE Region 10 Symposium (TENSYMP)*, 2017, pp. 1–5.
19. A. Maamar and K. Benahmed, "Machine learning Techniques for Energy Theft Detection in AMI," in *Proceedings of the 2018 International Conference on Software Engineering and Information Management - ICSIM2018*, 2018, pp. 57–62.
20. F. Al-Turjman, "5G-enabled Devices and Smart-Spaces in Social-IoT: An Overview," *Elsevier Future Gener. Comput. Syst.*, vol. 92, no. 1, pp. 732–744, 2019.
21. R. Rashed Mohassel, A. Fung, F. Mohammadi, and K. Raahemifar, "A Survey on Advanced Metering Infrastructure," *Int. J. Electr. Power Energy Syst.*, vol. 63, pp. 473–484, Dec. 2014.
22. A. K. Chakraborty and N. Sharma, "Advanced metering infrastructure: Technology and challenges," in *2016 IEEE/PES Transmission and Distribution Conference and Exposition (T&D)*, 2016, pp. 1–5.
23. K. S. Weranga, S. Kumarawadu, and D. P. Chandima, *Smart Metering Design and Applications*. Singapore: Springer Singapore, 2014.
24. G. A. Ajenikoko and A. A. Olaomi, "Hardware Design of a Smart Meter," *J. Eng. Res. Appl. www.ijera.com ISSN*, vol. 4, no. 6, 2014.

25. R. Yu, Y. Zhang, S. Gjessing, C. Yuen, S. Xie, and M. Guizani, "Cognitive Radio Based Hierarchical Communications Infrastructure for Smart Grid," *IEEE Netw.*, vol. 25, no. 5, pp. 6–14, Sep. 2011.

26. D. Baimel, S. Tapuchi, and N. Baimel, "Smart grid communication technologies- overview, research challenges and opportunities," in *2016 International Symposium on Power Electronics, Electrical Drives, Automation and Motion (SPEEDAM)*, 2016, pp. 116–120.

27. F. Al-Turjman, and A. Abdulsalam, "Smart-Grid and Solar Energy Harvesting in the IoT Era: An Overview," *Wiley's Concurrency and Computation: Practice and Experience,* 2018. DOI. 10.1002/cpe.4896.

28. L. Chhaya, P. Sharma, A. Kumar, and G. Bhagwatikar, "Communication Theories and Protocols for Smart Grid Hierarchical Network," *J. Electr. Electron. Eng.*, vol. 10, no. 1, pp. 43–48, 2017.

29. A. Y. Mulla, J. J. Baviskar, F. S. Kazi, and S. R. Wagh, "Implementation of ZigBee/802.15.4 in Smart Grid communication and analysis of power consumption: A case study," in *2014 Annual IEEE India Conference (INDICON)*, 2014, pp. 1–7.

30. A. A. Cecilia and K. Sudarsanan, "A survey on smart grid," in *2016 International Conference on Emerging Trends in Engineering, Technology and Science (ICETETS)*, 2016, pp. 1–7.

31. N. Shaukat et al., "A Survey on Consumers Empowerment, Communication Technologies, and Renewable Generation Penetration within Smart Grid," *Renew. Sustain. Energy Rev.*, vol. 81, pp. 1453–1475, Jan. 2018.

32. F. Al-Turjman, E. Ever,and H. Zahmatkesh, "Small Cells in the Forthcoming 5G/IoT: Traffic Modelling and Deployment Overview," *IEEE Communications Surveys and Tutorials*, 2018. DOI. 10.1109/COMST.2018.2864779.

33. S. K. Viswanath et al., "System Design of the Internet of Things for Residential Smart Grid," *IEEE Wirel. Commun.*, vol. 23, no. 5, pp. 90–98, Oct. 2016.

34. R. Want, "Near Field Communication," *IEEE Pervasive Comput.,* vol. 10, no. 3, pp. 4–7, Jul./Sep. 2011.

35. A. Mahmood, N. Javaid, and S. Razzaq, "A Review of Wireless Communications for Smart Grid," *Renew. Sustain. Energy Rev.*, vol. 41, pp. 248–260, Jan. 2015.

36. J. de Carvalho Silva, J. J. P. C. Rodrigues, A. M. Alberti, P. Solic, and A. L. L. Aquino, "LoRaWAN - A low power WAN protocol for Internet of Things: A review and opportunities," *2017 2nd International Multidisciplinary Conference on Computer and Energy Science,* pp. 1–6, 2017.

37. H. G. S. Filho, J. P. Filho, and V. L. Moreli, "The adequacy of LoRaWAN on smart grids: A comparison with RF mesh technology," in *2016 IEEE International Smart Cities Conference (ISC2)*, 2016, pp. 1–6.

38. K. Mochizuki, K. Obata, K. Mizutani, and H. Harada, "Development and field experiment of wide area Wi-SUN system based on IEEE 802.15.4g," in *2016 IEEE 3rd World Forum on Internet of Things (WF-IoT)*, 2016, pp. 76–81.

39. R. S. Sinha, Y. Wei, and S.-H. Hwang, "A Survey on LPWA Technology: LoRa and NB-IoT," *ICT Express,* vol. 3, no. 1, pp. 14–21, 2017.

40. W. Wang, Y. Xu, and M. Khanna, "A Survey on the Communication Architectures in Smart Grid," *Comput. Networks*, vol. 55, no. 15, pp. 3604–3629, Oct. 2011.

41. G. Singh, and F. Al-Turjman, "Cognitive Routing for Information-Centric Sensor Networks in Smart Cities" in *Proceedings of the International Wireless Communications and Mobile Computing Conference (IWCMC)*, Nicosia, Cyprus, 2014, pp. 1124–1129.

42. W.-N. Hsieh and I. Gitman, "Routing Strategies in Computer Networks," *Computer (Long. Beach. Calif).*, vol. 17, no. 6, pp. 46–56, Jun. 1984.

43. P. Bell and K. Jabbour, "Review of Point-to-Point Network Routing Algorithms," *IEEE Commun. Mag.*, vol. 24, no. 1, pp. 34–38, Jan. 1986.

44. R. Hou, C. Wang, Q. Zhu, and J. Li, "Interference-Aware QoS Multicast Routing for Smart Grid," *Ad Hoc Networks*, vol. 22, pp. 13–26, Nov. 2014.

45. Q. Wang and F. Granelli, "An improved routing algorithm for wireless path selection for the smart grid distribution network," in *2014 IEEE International Energy Conference (ENERGYCON)*, 2014, pp. 800–804.

46. L. Zhang, X. Liu, Y. Zhou, and D. Xu, "A Novel Routing Algorithm for Power Line Communication over a Low-voltage Distribution Network in a Smart Grid," *Energies*, vol. 6, no. 3, pp. 1421–1438, Mar. 2013.

47. S. Chaudhry, H. Alhakami, A. Baz, F. Al-Turjman, "Securing Demand Response Management: A Certificate based Authentication Scheme for Smart Grid Access Control," *IEEE Access*, vol. 8, no. 1, pp. 101235–101243, 2020.

48. F. Al-Turjman and K. Killic, "LaGOON: A Simple Energy-Aware Routing Protocol for Wireless Nano Sensor Networks," *IET Wireless Sensor Systems*, 2019. DOI: 10.1049/iet-wss.2018.5079.

49. A. Noorwali, R. Rao, and A. Shami, "End-to-end delay analysis of Wireless Mesh Backbone Network in a smart grid," in *2016 IEEE Canadian Conference on Electrical and Computer Engineering (CCECE)*, 2016, pp. 1–6.

50. W. Sun, J. P. Wang, J. L. Wang, Q. Y. Li, and D. M. Mu, "QoS Routing Algorithm of WSN for Smart Distribution Grid," *Adv. Mater. Res.*, vol. 1079–1080, pp. 724–729, Dec. 2014.

51. F. Al-Turjman, "Modelling Green Femtocells in Smart-grids," *Springer Mobile Networks and Applications*, vol. 23, no. 4, pp. 940–955, 2018.

52. J. Guo, J. Yao, T. Song, J. Hu, and M. Liu, "A routing algorithm to long lifetime network for the intelligent power distribution network in smart grid," in *2015 IEEE Advanced Information Technology, Electronic and Automation Control Conference (IAEAC)*, 2015, pp. 1077–1082.

53. X. Li, Q. Liang, and F. C. Lau, "A Maximum Likelihood Routing Algorithm for Smart Grid Wireless Network," *EURASIP J. Wirel. Commun. Netw.*, vol. 2014, no. 1, p. 75, Dec. 2014.

54. R. Wang, J. Wu, Z. Qian, Z. Lin, and X. He, "A Graph Theory Based Energy Routing Algorithm in Energy Local Area Network," *IEEE Trans. Ind. Informatics*, vol. 13, no. 6, pp. 3275–3285, Dec. 2017.

55. R. Rastgoo and V. S. Naeini, "A neurofuzzy QoS-aware routing protocol for smart grids," in *2014 22nd Iranian Conference on Electrical Engineering (ICEE)*, 2014, pp. 1080–1084.

56. F. Al-Turjman and S. Alturjman, "Confidential Smart-Sensing Framework in the IoT Era," *The Springer Journal of Supercomputing*, vol. 74, no. 10, pp. 5187–5198, 2018.

57. Y. Zhang, W. Sun, and L. Wang, "Location and communication routing optimization of trust nodes in smart grid network infrastructure," in *2012 IEEE Power and Energy Society General Meeting*, 2012, pp. 1–8.

58. F. Al-Turjman, "The Road Towards Plant Phenotyping via WSNs: An Overview," *Elsevier Computers & Electronics in Agriculture*, 2018. DOI: 10.1016/j.compag.2018.09.018.

59. L. Wisniewski, "Communication Reliability," in *New Methods to Engineer and Seamlessly Reconfigure Time Triggered Ethernet Based Systems During Runtime Based on the PROFINET IRT Example*, Berlin, Heidelberg: Springer Berlin Heidelberg, 2017, pp. 105–123.

60. P. Zhao, P. Yu, C. Ji, L. Feng, and W. Li, "A routing optimization method based on risk prediction for communication services in smart grid," in *2016 12th International Conference on Network and Service Management (CNSM)*, 2016, pp. 377–382.

61. P. Rohal, R. Dahiya, and P. Dahiya, "Study and analysis of throughput, delay and packet delivery ratio in MANET for topology based routing protocols (AODV, DSR and DSDV)," *www.ijaret.org Issue II*, vol. 1, 2013.

62. Y. Zong, Z. Zheng, and M. Huo, "Improving the reliability of HWMP for smart grid neighborhood area networks," in *2016 International Conference on Smart Grid and Clean Energy Technologies (ICSGCE)*, 2016, pp. 24–30.

63. Y. Qian, C. Zhang, Z. Xu, F. Shu, L. Dong, and J. Li, "A Reliable Opportunistic Routing for Smart Grid with In-Home Power Line Communication Networks," *Sci. China Inf. Sci.*, vol. 59, no. 12, p. 122305, Dec. 2016.

64. A. Kannan, V. Kumar, T. Chandrasekar, and B. J. Rabi, "A review of power quality standards, electrical software tools, issues and solutions," in *2013 International Conference on Renewable Energy and Sustainable Energy (ICRESE)*, 2013, pp. 91–97.

65. J. A. Orr and B. A. Eisenstein, "Summary of Innovations in Electrical Engineering Curricula," *IEEE Trans. Educ.*, vol. 37, no. 2, pp. 131–135, May 1994.

66. J. S. Subjak and J. S. McQuilkin, "Harmonics-Causes, Effects, Measurements, and Analysis: An Update," *IEEE Trans. Ind. Appl.*, vol. 26, no. 6, pp. 1034–1042, 1990.

67. M. H. J. Bollen and I. Y. Gu, "Appendix B: IEEE Standards on Power Quality," in *Signal Processing of Power Quality Disturbances*, Hoboken, NJ: John Wiley & Sons, Inc., pp. 825–827.

68. M. H. J. Bollen and I. Y. Gu, "Appendix A: IEC Standards on Power Quality," in *Signal Processing of Power Quality Disturbances*, Hoboken, NJ: John Wiley & Sons, Inc., pp. 821–824.

69. A. Nayyar, F. Al-Turjman and L. Mostarda, "Proficient QoS Based Target Coverage Problem in Wireless Sensor Networks," *IEEE Access*, vol. 8, no. 1, pp. 74315–74325, 2020.

70. H. Zahmatkesh, F. Al-Turjman, "Fog Computing for Sustainable Smart Cities in the IoT Era: Caching Techniques and Enabling Technologies - An Overview", *Elsevier Sustainable Cities and Societies*, vol. 59, 102139, 2020.

71. F. Al-Turjman, and C. Altrjman, "Enhanced Medium Access for Traffic Management in Smart-cities' Vehicular-Cloud," *IEEE Intelligent Transportation Systems Magazine*, 2020. DOI. 10.1109/MITS.2019.2962144.

72. F. Al-Turjman, and C. Altrjman, S. Din, and A. Paul, "Energy Monitoring in IoT-based Ad Hoc Networks: An Overview", *Elsevier Computers & Electrical Engineering Journal*, vol. 76, pp. 133–142, 2019.

73. "IEC 61000-4-30 Ed3 - Power Standards Lab." [Online]. Available: https://www.powerstandards.com/testing-certification/certification-standards/iec-61000-4-30-ed3-certification-testing/. [Accessed: 02-Dec-2017].

74. C. Masetti, "Revision of European standard EN 50160 on power quality: Reasons and solutions," in *Proceedings of 14th International Conference on Harmonics and Quality of Power - ICHQP 2010*, 2010, pp. 1–7.

75. M. M. Albu, M. Sanduleac, and C. Stanescu, "Syncretic Use of Smart Meters for Power Quality Monitoring in Emerging Networks," *IEEE Trans. Smart Grid*, vol. 8, no. 1, pp. 485–492, Jan. 2017.

76. Power System Instrumentation and Measurements of the IEEE Power Engineering Society, "*IEEE Std 1459-2010, IEEE Standard Definitions for the Measurement of Electric Power Quantities Under Sinusoidal, Nonsinusoidal, Balanced, or Unbalanced Conditions.*"

77. L. Morales-Velazquez, R. de J. Romero-Troncoso, G. Herrera-Ruiz, D. Morinigo-Sotelo, and R. A. Osornio-Rios, "Smart Sensor Network for Power Quality Monitoring in Electrical Installations," *Measurement*, vol. 103, pp. 133–142, June 2017.

78. E. Ever, F. Al-Turjman, H. Zahmatkesh, and M. Riza, "Modelling Green HetNets in Presence of Failures for Dynamic Large-Scale Applications: A Case-study for Fault Tolerant Femtocells in Smart cities," *Elsevier Computer Networks Journal*, vol. 128, pp. 78–93, 2018.

79. P. Koponen, R. Seesvuori, and R. Bostman, "Adding Power Quality Monitoring to a Smart kWh Motor," *Power Eng. J.*, vol. 10, no. 4, pp. 159–163, Aug. 1996.
80. K. D. McBee and M. G. Simoes, "Utilizing a Smart Grid Monitoring System to Improve Voltage Quality of Customers," *IEEE Trans. Smart Grid*, vol. 3, no. 2, pp. 738–743, June 2012.
81. F. A. S. Borges, R. A. S. Fernandes, I. N. Silva, and C. B. S. Silva, "Feature Extraction and Power Quality Disturbances Classification Using Smart Meters Signals," *IEEE Trans. Ind. Informatics*, vol. 12, no. 2, pp. 824–833, Apr. 2016.
82. W. G. Morsi, "Electronic Reactive Energy Meters' Performance Evaluation in Environment Contaminated with Power Quality Disturbances," *Electr. Power Syst. Res.*, vol. 84, no. 1, pp. 201–205, Mar. 2012.
83. C. De Capua and E. Romeo, "A Smart THD Meter Performing an Original Uncertainty Evaluation Procedure," *IEEE Trans. Instrum. Meas.*, vol. 56, no. 4, pp. 1257–1264, Aug. 2007.
84. F. G. Montoya, A. García-Cruz, M. G. Montoya, and F. Manzano-Agugliaro, "Power Quality Techniques Research Worldwide: A Review," *Renew. Sustain. Energy Rev.*, vol. 54, pp. 846–856, Feb. 2016.
85. D. Granados-Lieberman, R. J. Romero-Troncoso, R. A. Osornio-Rios, A. Garcia-Perez, and E. Cabal-Yepez, "Techniques and Methodologies for Power Quality Analysis and Disturbances Classification in Power Systems: A Review," *IET Gener. Transm. Distrib.*, vol. 5, no. 4, p. 519, 2011.
86. S. Chen and H. Y. Zhu, "Wavelet Transform for Processing Power Quality Disturbances," *EURASIP J. Adv. Signal Process.*, vol. 2007, no. 1, p. 047695, Dec. 2007.
87. N. ul Hasan, W. Ejaz, M. Atiq, and H. Kim, "Recursive Pyramid Algorithm-Based Discrete Wavelet Transform for Reactive Power Measurement in Smart Meters," *Energies*, vol. 6, no. 9, pp. 4721–4738, Sep. 2013.
88. N. C. F. Tse, J. Y. C. Chan, W.-H. Lau, J. T. Y. Poon, and L. L. Lai, "Real-Time Power-Quality Monitoring With Hybrid Sinusoidal and Lifting Wavelet Compression Algorithm," *IEEE Trans. Power Deliv.*, vol. 27, no. 4, pp. 1718–1726, Oct. 2012.
89. E. Junput, S. Chantree, M. Leelajindakrairerk, and C.-C. Chompoo-inwai, "Optimal technique for total harmonic distortion detection and estimation for smart meter," in *2012 10th International Power & Energy Conference (IPEC)*, 2012, pp. 369–373.
90. G. T. Heydt, P. S. Fjeld, C. C. Liu, D. Pierce, L. Tu, and G. Hensley, "Applications of the Windowed FFT to Electric Power Quality Assessment," *IEEE Trans. Power Deliv.*, vol. 14, no. 4, pp. 1411–1416, 1999.
91. A. Augustine et al., "Review of Signal Processing Techniques for Detection of Power Quality Events," *Am. J. Eng. Appl. Sci.*, vol. 9, no. 2, pp. 364–370, Feb. 2016.
92. S. Alabady and F. Al-Turjman, "Low Complexity Parity Check Code for Futuristic Wireless Networks Applications," *IEEE Access J*, vol. 6, no. 1, pp. 18398–18407, 2018.
93. R. Billinton and R. N. Allan, *Reliability Evaluation of Power Systems*. Boston, MA: Springer US, 1996.
94. K. Kuhi, K. Korbe, O. Koppel, and I. Palu, "Calculating power distribution system reliability indexes from Smart Meter data," in *2016 IEEE International Energy Conference (ENERGYCON)*, 2016, pp. 1–5.
95. IEEE Power & Energy Society. Transmission and Distribution Committee., Institute of Electrical and Electronics Engineers., and IEEE-SA Standards Board., IEEE guide for electric power distribution reliability indices. Institute of Electrical and Electronics Engineers, 2012.
96. O. Siirto, M. Hyvärinen, M. Loukkalahti, A. Hämäläinen, and M. Lehtonen, "Improving Reliability in an Urban Network," *Electr. Power Syst. Res.*, vol. 120, pp. 47–55, Mar. 2015.

97. J. R. Hammarstron, A. da R. Abaide, M. W. Fuhrmann, and E. A. L. Vianna, "The impact of the installation of smart meters on distribution system reliability," in *2016 51st International Universities Power Engineering Conference (UPEC)*, 2016, pp. 1–5.

98. S.-W. Luan, J.-H. Teng, S.-Y. Chan, and L.-C. Hwang, "Development of an automatic reliability calculation system for advanced metering infrastructure," in *2010 8th IEEE International Conference on Industrial Informatics*, 2010, pp. 342–347.

99. C. Gamroth, Kui Wu, and D. Marinakis, "A smart meter based approach to power reliability index for enterprise-level power grid," in *2012 IEEE Third International Conference on Smart Grid Communications (SmartGridComm)*, 2012, pp. 534–539.

100. A. Mohsenzadeh, S. Pazouki, M.-R. Haghifam, and M. E. Talebian, "Impact of DLC programs levels on reliability improvement of smart distribution network considering multi carrier energy networks," in *2013 8th International Conference on Electrical and Electronics Engineering (ELECO)*, 2013, pp. 72–76.

2 Energy Monitoring in IoT-Based Grid

Fadi Al-Turjman
Near East University

Chadi Altrjman
University of Waterloo

Sadia Din and Anand Paul
Kyungpook National University

CONTENTS

2.1   INTRODUCTION

Smart spaces are solutions where the Internet of Things (IoT)-enabling technologies have been employed towards further advances in the life style. It tightly integrates with the existing cloud infrastructure to impact several fields in the academia and/or industry. IoT has been recently recognized as a disruptive technology for the flying ad hoc network. It can be viewed as the network of networks. IoT-based networks are communication networks with users moving with low speed (walking speed) to vehicular, train or plane speed, and hinge largely upon the so-called self-organizing capability operating in a distributed fashion in ad hoc networks. There can be a wide range of applications in IoT that supports logistics and ad hoc network management. IoT technology can be leveraged to achieve cost reductions. Moreover, this technology can be combined with real-time localization systems to get live updates from the factory floor, to enable manufacturers to continuously monitor machine activity, to satisfy maintenance needs, and to produce movement during production. Cost reduction can be achieved across the digital ad hoc networks by making use of these smart machines while providing data that allow manufacturers to adjust production on the fly. Manufacturing and assembly lines will receive updated schedules and quality-related information in real time and instantaneously. In addition, IoT data can be leveraged to schedule proactive, preventive, and predictive repairs and maintenance. It can also be utilized to customize production to meet the customer's orders and the focus that is needed to be successful in the digital world. The concept of Industry 4.0 will soon be a reality as well while utilizing energy-aware IoT paradigms. Smart products consisting of the embedded knowledge of the customers can also provide data insights and analytics to achieve the best customer experience. All the aforementioned efforts can lead to more cost-efficient energy production and development.

The utilization of IoT in specially appointed systems such as the ad hoc networks of things is expanding quickly. Deutsche Bahn, the German railways and freight bearer, introduced a system-wide checking framework to deal with its whole rail network. This network includes more than 1 billion supply chain "hubs," which is gathering information on each section of track, rail vehicle, station, motor, and switch, and checking the state of these things in an ongoing way. The gathered information is bolstered into a control tower that totals them periodically to give close ongoing data over the whole armada. Deutsche Bahn has utilized this information to enhance hazard practices such as ongoing rerouting and optimization, considering all current system traffic bypassing through these hubs. Whirlpool is another case of utilizing the IoT for interior supply chain vitality enhancement in directing work and finding lost stock. Rather than utilizing standardized bar codes or a comparable arrangement, Whirlpool utilized radio frequency identification (RFID) labels and readers over an assembling plant to give chiefs and administrators constant access to data for inbound coordination to the paint line. The IoT can be used by the various partners of the supply chain as well to monitor its execution process in real time and improve the efficiency and effectiveness of the energy consumption and savings. Recent developments in the IoT have made it possible to achieve high visibility in production networks. For example, the IoT benefits the food and agricultural products by improving their visualization and traceability and by assuring people's food safety. Industrial deployment of the IoT provides development of an ideal platform for decentralized management of warehouses and collaborative warehouse order fulfillment using RFID, ambient intelligence, and multi-agent system.

Due to the wide use of IoT paradigm nowadays, the energy consumption factor is rising as a key issue in the forthcoming age of the associated telecommunication systems. Therefore, it is worth mentioning the main factors which contribute towards more energy consumption in IoT. Key energy consumers in the IoT era can be appliances in smart homes/smart buildings, sensor ad hoc networks in traffic and health systems, RFID tags/readers in smart identifiable spaces, etc.

Notwithstanding the huge enthusiasm by IoT and production networks' managers and other applications, there is yet a lack in energy consumption monitoring and optimization. Observing and monitoring frameworks in this domain can gather and send key information about the monitored resource condition, hardware execution, testing, vitality utilization, and natural conditions, and enable administrators and mechanized controllers to react to changes progressively anywhere. These capacities are vital to keen production network exercises where deceivability and traceability of products are required.

RFID is a piece of the IoT frameworks identified with inventory network exercises. RFID systems utilize radio frequency waves, a tag, and a reader. The tag can store a bigger amount of information than customary scanner tags. Meanwhile, RFID readers can recognize, track, and screen the items connected with labels all around, naturally and progressively, if necessary. Nonetheless, remote advancements have assumed a key part in mechanical observing and control frameworks. In an average production network, observing and control applications utilize sensors, GPS, RFID labels, and sensor systems to limit scattering, robbery, misfortune, and decay in distribution center, transportation, and store racks. Sensors are utilized to keep up merchandise at the right temperature and shield them from substance spills to waste. Sensor systems screen movement conditions, route gadgets, and track the area of transportation vehicles to make steering more proficient. The wireless sensor network (WSN) comprises spatially dispersed self-sufficient sensor-prepared gadgets to screen physical or ecological conditions and can participate with RFID frameworks to better track the status of things such as their area, temperature, and developments. WSN has been principally utilized as a part of frosty chain coordinations which includes the transportation of temperature delicate items along an inventory network through warm and refrigerated bundling strategies. It is additionally utilized for up keeping and following frameworks. General electric (GE) devices convey sensors in its motors, turbines, and wind ranches. By dissecting information continuously, GE spares time and expenses because of the convenient preventive support.

Energy consumed by the aforementioned IoT-enabling technologies and appliances can be characterized into online versus passive energy consumers. In the online (active) mode, the utilized enabling technology is using energy via a continuous electric source such as the wall electricity or the gas pipe. In the offline (passive) one, energy is used based on its availability such as those photonic sensors which rely on solar panels in harvesting their energy from the surrounding environment. In both cases, it is worth mentioning that monitoring the energy usage can save the service provider a lot in terms of the operational cost and the performance efficiency. Accordingly, monitoring the cumulative amount of used Watts per hour in the aforementioned modes plays a significant role in any energy-monitoring system. Therefore, energy providers have a great interest nowadays in installing the watt hour meters (WHM) everywhere. The communication between the utilities and the consumer in production networks (supply chain) helps a lot in terms of energy conservation. The behaviors of the consumer vary with the changes in the utility pricing, but it is also affected if the consumer is aware of the appliance's consumption. This kind of a direct feedback helps in conserving energy as well as managing the resources for both customer and supplier.

To comprehend the significance of energy checking, we take a use case of a petrol store, where all individuals from one family go to refill their vehicle's fuel tanks yet nobody pays in the wake of refilling since they have a month-to-month charging framework with the service station proprietor. Towards the end of the month when the family gets the bill, they do not have the foggiest idea who they should represent such a huge bill. Similarly, getting the utility/electricity bill toward the end of the month without knowing which apparatus/appliance is contributing the amount to the bill. Therefore, it becomes difficult to monitor the vitality. This is where direct vitality checking helps in giving a superior arrangement to use energy proficiently while saving by observing where the vitality is being utilized and what should be possible to lessen the utilization.

Smart energy monitoring (SEM) works on a very basic principle where the energy being supplied by the provider company enters the home by WHM from where it goes to load survey meter (LSM) which calculates the total load of the entire building. Coming down in the system are end user meters (EUM) installed separately for all the appliances which calculate the energy consumption for individual appliances. Finally, all the data from the end-user meters is collected and distributed to monitor servers as depicted in Figure 2.1.

The informational terminal, regardless of the type of monitoring system that the user has installed, comprises a detailed logging of the energy consumption of the entire unit/building. It illustrates the electricity charges, daily loads, daily changes in consumption over last few days/week, and comparison between past data for both individual appliances and entire building along with energy conservation tips as shown in Figure 2.2.

Energy conservation is a big challenge for both customer and supplier, but the beneficiaries of this conservation are again for both parties. The consumers who

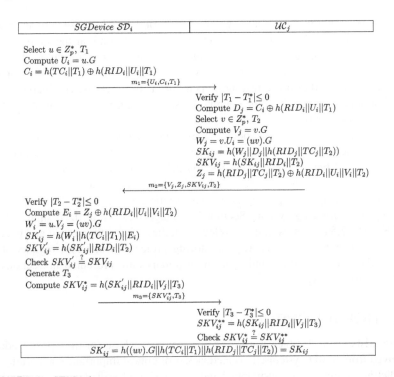

SGDevice $\mathcal{SD}_i$	$\mathcal{UC}_j$

Select $u \in Z_p^*$, T_1
Compute $U_i = u.G$
$C_i = h(TC_i||T_1) \oplus h(RID_i||U_i||T_1)$

$\xrightarrow{\quad m_1 = \{U_i, C_i, T_1\} \quad}$

Verify $|T_1 - T_1^*| \leq 0$
Compute $D_j = C_i \oplus h(RID_i||U_i||T_1)$
Select $v \in Z_p^*$, T_2
Compute $V_j = v.G$
$W_j = v.U_i = (uv).G$
$SK_{ij} = h(W_j||D_j||h(RID_j||TC_j||T_2))$
$SKV_{ij} = h(SK_{ij}||RID_i||T_2)$
$Z_j = h(RID_j||TC_j||T_2) \oplus h(RID_i||U_i||V_i||T_2)$

$\xleftarrow{\quad m_2 = \{V_j, Z_j, SKV_{ij}, T_2\} \quad}$

Verify $|T_2 - T_2^*| \leq 0$
Compute $E_i = Z_j \oplus h(RID_i||U_i||V_i||T_2)$
$W_i' = u.V_j = (uv).G$
$SK_{ij}' = h(W_i'||h(TC_i||T_1)||E_i)$
$SKV_{ij}' = h(SK_{ij}'||RID_i||T_2)$
Check $SKV_{ij}' \overset{?}{=} SKV_{ij}$
Generate T_3
Compute $SKV_{ij}^* = h(SK_{ij}'||RID_i||V_j||T_3)$

$\xrightarrow{\quad m_3 = \{SKV_{ij}^*, T_3\} \quad}$

Verify $|T_3 - T_3^*| \leq 0$
$SKV_{ij}^{**} = h(SK_{ij}||RID_i||V_j||T_3)$
Check $SKV_{ij}^* \overset{?}{=} SKV_{ij}^{**}$

$SK_{ij}' = h((uv).G||h(TC_i||T_1)||h(RID_j||TC_j||T_2)) = SK_{ij}$

FIGURE 2.1   SEM scheme.

Electricity Charge of TV set 15.92 USD		Appliances
		Gas Heater
Comparison with yesterday 6% higher		Air Conditioner 1
		Air Conditioner 2
Yesterday Electricity Charge 15.01 USD		Electric Heater 1
		Electric Heater 2
Charges of this Month 132 USD		TV Set
		Microwave
Choose Period		Dish washer
Daily	10 Days	Refrigerator
Compare with past date		Washing Machine
By Appliances	By Dates	Desk PC
By Month	By Year	Whole house
11 December 2017	User Tips	Percentage of Each Appliance
In-Home Information Terminal		

FIGURE 2.2   In-home information display terminal.

participate in energy conservation get the benefit of direct decrease in their electricity bills as well as less carbon footprint. On the other hand, at the supplier level, conservation of energy means a continuous monitoring on consumption growth which may yield economic benefits by saving the expenditures on extra capacity investments as well as peak shifting. This in turn helps to ensure continuous supply. In addition, it may also help the country for saving the GHG emissions and relying on renewable resources.

A detailed review study about the existing energy-monitoring projects and methodologies is conducted in this chapter to compare different design aspects and to recommend better alternatives. Section 2.2 categorizes and discusses energy-monitoring systems based on feedback and effectiveness, and also subcategorizes them based on type of pricing strategies. Section 2.3 lists the devices used for sensing in monitoring system. Section 2.4 categorizes the utilized smart spaces, and Section 2.5 discusses in detail relevant studies related to the field and presents promising results of pilot energy-monitoring projects in different regions of the world. Further discussions and concluding remarks are presented in Sections 2.6 and 2.7, respectively.

2.2 ENERGY MONITORING

There exists a considerable variety in energy monitoring based on the type of feedback received from power systems. This feedback is of utmost importance in learning and predicting future energy consumption and savings. It is classified into two types, indirect feedback systems and direct feedback systems, which are discussed next.

2.1.1 INDIRECT FEEDBACK SYSTEMS

Indirect feedback systems are usually managed by the energy/service supplier (utility companies) where the raw data collected is administered and later on sent to clients with their end of the month bills. In this system, several cloud services can be used offline in order to provide the feedback to users. Cloud enabled with IoT features integrates smoothly with the emerging 5G networks in order to provide this indirect feedback. The greatest effectiveness out of this monitoring type can be accomplished by revising these bills dependent on their authentic, similar criticisms and periodic vitality remarks. Nevertheless, this kind of monitoring cannot be applied in real-time manners/opportunities. This can dramatically affect the system accuracy in predicting and reporting the most effective solutions.

2.1.2 DIRECT FEEDBACK SYSTEMS

Direct feedback systems perform a real-time energy monitoring with the help of the cloud-enabled IoT spaces, which includes smart meters (SMs) and in-home displays (IHDs) to visualize the power consumption information, in addition to charging rates and usages. Big data paradigms are used to analyze the different behaviors experienced by different types of appliances. Direct feedback systems can further be classified based on the utilized pricing strategy as follows: (1) IHD in which the user is provided with an IHD unit for a monthly billing system such that the price of energy consumption is time independent and fixed for entire month, and (2) pay as you go is a pricing policy in which the users of electricity are entitled to buy the electricity prior to consumption using a smart card which can be recharged at the utility provider side. Users can plug in this smart card to activate their energy SMs. Once this card runs out of the permitted energy amount, the SM cuts the electricity connection or sends a warning message. (3) Time of use (TOU) is another pricing method in which the price per energy unit is time dependent. This means that the price of energy consumption per unit can be higher in peak hours and lower in off-peak hours. This allows users of TOU to be aware of the used energy price per unit during peak and/or off-peak hours.

2.3   ENERGY-MONITORING DEVICES

Essential component of any energy-monitoring system is the sensing device. In order to monitor energy consumption using these sensing devises, two critical parameters must be known: voltage and current. In order to measure these parameters, different methods have been used while using different types of appliances. These methods can be categorized into direct and indirect sensing approaches, which are discussed next.

2.3.1   DIRECT SENSING

Direct sensing is the most commonly used method in which the electric current and voltage used by the appliance is measured directly. In this method, the current in the appliance is measured in series using an ammeter, whereas the voltage across the appliance is measured in parallel using a voltmeter. Since these parameters are to be measured at all used appliances, large-scale economic devices are required. In order to overcome this cost issue, a microprocessor can be embedded and programed to control a chain of these sensing devices. In the following, we list a few examples of this category.

2.3.1.1   ILD Sensors

Inductive loop detector (ILD) is normally installed on walls in circular or rectangular shapes. They consist of several wire loops where electric current is passing through them generating an electromagnetic field with inductance, and while in presence or passing of an object, they excite a frequency between 10 and 50 kHz resulting in a reduction in the inductance that causes the electronic unit to oscillate with higher frequency. This ultimately sends a pulse to the controller indicating the passing of an energy supply unit.

2.3.1.2   RFID

RFID can be used for the object passing detection in many related applications. RFID systems consist of readers and tags. RFID tag or transponder with its unique ID can be read via the reader or a transponder antenna. It can be placed inside the appliances and the reader antenna that is placed in the building, can read the tag, and can change the occupancy/presence of the appliance. With this system, the delay can be minimized, and the flow of the data traffic can be achieved in a smoother way.

2.3.1.3   Infrared Sensors

Infrared sensors work by detecting the temperature difference of an area/object in relevant applications. It identifies the usage of an appliance by measuring the temperature difference in the form of thermal energy emitted by the appliance and the surrounding environment. Infrared sensors, unlike the other types, do not require to be anchored or tunneled into the ground or the wall, but rather they are mounted to the ceiling or the ground. However, they are prone to weather conditions effects that can degrade the performance sometimes.

2.3.2   Indirect Sensing

Indirect sensing devices are more appropriate for in-house monitoring systems, where energy interference engines are implemented using magnetic, photonic, and/or acoustic sensors to measure the emitted energy from the used appliances. This type of devices is usually integrated in wireless radio-enabled networks in order to aggregate and forward the collected data. Furthermore, it supports energy-aware algorithms which can be run in a distributed fashion for power consumption/prediction at each house/building. Commonly used sensing devices under this category can be the Crossbow MicaZ motes which supports different sensing methods as detailed next.

2.3.2.1   Magnetic Sensors

One of the commonly used magnetic sensors is the HMC 1002. It is a surface-mounted sensor that is designed for low-field magnetic sensing as well as navigation systems. It has magnetometers and current sensing elements that make use of anisotropic magneto-resistive (AMR) technology, and it outperforms coil-based sensors. It senses the changes in the magnetic field. There are two types of magnetometers: single-axis and double-/triple-axis magnetometer, wherein the latter case, the accuracy of changes detection is much higher due to the fact that it uses two/three axes. This magnetic sensor measures the magnetic field changes around the input power cable of the appliance and reports the amount of energy used [1].

2.3.2.2   Acoustic Sensors

Acoustic sensors can detect energy that is produced by appliances or their interactions. In the detection zone of the sensor, a single algorithm can detect and signal the presence of a functioning device. It is used to sense the internal power stats of an appliance in terms of noise pattern. Likewise, in the drop of the sound level, the appliance usage is considered terminated.

2.3.2.3   Light and Piezoelectric Sensors

There are several light sensors for monitoring systems in the literature. The most commonly used sensors are CdSe photocell light intensity sensors and simple piezoelectric sensors [2]. Piezoelectric sensors detect mechanical stress that is induced by pressure or vibration of surrounding objects by converting it into an electrical signal. The value of the generated voltage is directly proportional to the weight of the object exerting a force on the sensors. For accurate measurements, multiple sensors should be used. However, they are susceptible to high amount of pressures and temperature. These light sensors are used to sense the internal power stats of an appliance in terms of light intensity.

2.4 SMART SPACES

In this section, we elaborate on the smart spaces, where smart monitoring systems and smart devices are integrated to achieve the most energy savings. Smart spaces take into account usually users' identities, location, and context relevant to the ongoing activity. It involves four key entities: the user, monitoring device, cloud, and the service provider. We categorize these spaces into online spaces and offline spaces, which are discussed next.

2.4.1 ONLINE SPACES

Online spaces necessitate a stable and continuous Internet connection to communicate data with a cloud-based service. These services are usually provided through a mobile app nowadays which can predict the consumed energy. Their prediction algorithm is designed to be online to predict the subsequent app needs of a user according to their geo-locations and app setups. However, it was found that most of the users turn off their Wi-Fi connection to save their battery charge. This can dramatically affect the monitoring systems due to unexpected outages.

2.4.2 OFFLINE SPACES

In this type, data is locally stored at the user end device and then periodically uploaded to centralized/distributed computing systems for processing. The authors proposed a method for predicting the app usage using check points and location information. They used semantic locations approach and created a database containing application launches. Their prediction algorithm is designed to be online only when it is connected to the cloud. It provides the analyzed results from the cloud and helps the user taking smarter decisions.

In Ref. [27], the authors proposed an integrated toolbox which has been designed to assist energy distributers in monitoring the energy usage of buildings in a systematic way. Online and offline energy monitoring and diagnosing was implemented using WSNs in Ref. [28]. Although these approaches have been verified through a series of laboratory experiments, practical issues have been noticed and experienced in the field. This is due to unpredicted WSN dynamics, noise and interference issues, and unexpected surrounding environmental conditions, which can significantly affect the wireless communication quality and reliability of the reported results.

2.5   USE CASES IN PRACTICE

Renewable energy resources are intermittent by nature. The increased penetration of those non-dispatchable energy sources into the existing power infrastructure makes it more challenging for utilities to deliver reliable and good quality power. Fortunately, with the recent technological advancements, IoT/cloud-inspired applications can offer great solutions to the aforementioned challenges by providing two-way communication schemes which can help in transforming conventional power infrastructure into modernized smart one through intelligent monitoring systems. These systems support a lot the emerging smart grid project in smart cities. They provide better monitoring for the status of distributed power to consumers. The functionalities of smart grids vary depending on the monitoring application objectives including energy demands saving, feedback delivery to consumers, and dynamic energy pricing. On the other hand, they control the appliances functionality based on demands curves, security enhancement, outage management, supply quality assessment, and the customer response management. Hence, smart grid can be the ultimate solution to most of the challenges in current power grids. In fact, statistical studies show that the component failures in a power system can cause more than 80% of the electricity outages/cuts in a power distribution grid. Hence, the smart grid can overcome the aforementioned challenges, when it is properly integrated with efficient monitoring systems.

Several studies have been conducted to evaluate the impact and effectiveness of energy-monitoring systems on consumer's behavior. The studies have consistently suggested that the consumer's response to the monitoring system was always positive in terms of energy consumption and savings. In addition, most studies have confirmed a reduction of up to 25% in the consumers' energy consumption. The geographical locations where these monitoring projects were implemented as well as the demographics of the consumers have been reported and varied from one place to another. This was due to the difference in the communal behavior as well as the climate variations. A number of pilot monitoring projects are explored, and the results from these projects have been analyzed and compared in this study.

2.5.1   Pilot Projects Using IHD

In this section, we look at the different pilot projects utilized worldwide in which an IHD is used to gauge the impact on the consumer's behavior.

2.5.1.1.   Hydro-One Project

The Hydro-One networks in the province of Ontario, Canada [3], started a pilot project in 2004 in which around 400 participants participated from different Canadian cities including Barrie, Brampton, Lincoln, Peterborough, and Timmins. The consumption patterns of the participants were monitored over a period of 2.5 years. After that, the participants were provided with an IHD system to monitor their consumption trends and cost savings. The impact of the IHD system was observed by Hydro-One. The real-time feedback provided by IHD resulted in an average decrement of 6.5% between all the subscribed consumers. However, the main limitation in this project was its complete dependability on the user in choosing the best time to switch on and off their appliances instead of using an automated system that avoids the peak times. The automated system can be implemented by deploying a simple chip to perform the task and result in a much more cost savings.

2.5.1.2   CEATI Project

Similar to the Hydro-One project, Customer Energy Solution Interest Group (CEATI) in Canada consisting of British Columbia (BC) Hydro, Newfoundland Power, National Rural Electric Cooperative Association, and Natural Resource Canada Office of Energy Efficiency undertook a project in the province of BC, Newfoundland and Labrador. In this project, a group of almost 200 users of Newfoundland Power and BC Hydro Consumers were convinced to participate knowing that no rate incentives will be given. This was one of the main limitations in this project. Statistical results of the project showed almost 18% average decrease in consumption of the participants from Newfoundland Power and 2.7% decrement in consumption of the BC Hydro consumers.

2.5.1.3   Cost Monitoring Project

In 2007, three major suppliers, in the state of Massachusetts, USA [5], namely, National Grid, NSTAR and Western Massachusetts Electric Company, conducted a similar project as described in the earlier projects by distributing thousands of the IHDs. The project mainly focused on cost saving analysis and consumers' opinions on their savings. The published results showed that half of the consumers saved around 5%–10% on their billings. However, the main limitation in delegating the final decision to the end user was still there.

2.5.1.4   SDG&E Project

In 2007, San Diego in the state of California, USA, started an energy-monitoring project with 300 participants [6, 7]. IHDs were installed for the residents whose monthly consumption was more than 700 kWh. The difference in this project compared to the one mentioned before was the phone calls and email notification for their energy consumption trend. Therefore, it was more dependent on the end-user feedback rather than a systematic way in computing.

2.5.1.5   ECOIS Project

In 2000, Osaka University launched a project called Energy Consumption Information System (ECOIS). In this project, the energy consumption was monitored for nine households over a time period of two years. After that, IHD was installed in the households [8]. The system was monitored by the servers for providing consumption notifications. Figure 2.3 demonstrates the trends of energy consumption before and after the installation of ECOIS system. It is evident from the figure that the energy consumption was reduced significantly after consumers started monitoring their energy consumption.

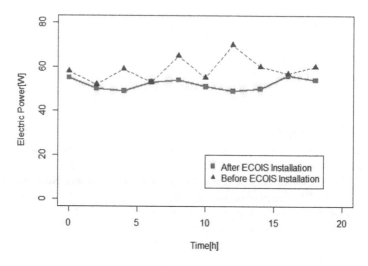

FIGURE 2.3   Load curve before and after installation of IHDs.

2.5.2   Projects with IHD and Prepay Method

The second category in the field of energy monitoring is the consumer-engaged pilot program using IHD. However, in this category, consumers are required to purchase the electricity prior to being consumed/used. Again, smart cards are used to purchase the electricity, and then, they are plugged inside an IHD system or inside an energy SM to start using the electricity. The followings are a few use cases under this category.

2.5.2.1   M-Power Project

In the state of Arizona, USA, a project named "Salt River" was launched. A total of 100 participants were provided with IHD systems as well as prepay meters to allow the use of smart cards to purchase electricity [9]. An average of 12.8% reduction was observed in the energy consumption. The M-Project is still active as one of the biggest programs in North America with more than 50,000 subscribers [10,11]. This one is more automated than the previous ones; however, it is considered as a small-scale project and lacks scalability in huge modern cities.

2.5.2.2   Woodstock Hydro

In the province of Ontario, Canada, the city of Woodstock launched a program in which 2500 participants voluntarily agreed to be a part of a pilot program [12]. The results gathered included the effect of both IHD and pay as you go being implemented together. This made it difficult to conclude which of the two methods conducted more energy savings. In general, this kind of monitoring projects is mainly dependent on the amount of money (expenses) as an indicator and as a metric to measure the amount of the used/consumed energy. This might not be that much accurate as energy should be measured by the watt-hour unit.

2.5.3 PROJECTS WITH INTEGRATED IHD/TOU

This is a third category in the field of SEM usually known as TOU in which the users are entitled to a time-varying rate according to the time of used energy. Usually, the rates are set at a higher price at the peak hours to reduce the demand, whereas they are set at lower price point for off-peak hours. Often, incentives are also given to promote less consumption during the peak hours. The following are a few use cases under this category.

2.5.3.1 Hydro-One TOU Project

Hydro-One networks in the province of Ontario, Canada, started a pilot project with 486 participants to implement TOU pricing policy along with IHD [13]. The group with both IHD and TOU consumed 7.6% less energy, whereas the group with TOU only reduced their energy usage by 3.3%. The group using IHD without TOU was able to consume 6.7% less energy. This suggests that using IHD alone was more effective than using TOU alone. In addition, it is clear that when the TOU pricing policy was introduced along with IHD, the energy consumption was reduced at a much significant rate.

2.5.3.2 Information Display Project

The Information Display Project of the USA in the state of California was experimented over a group of 61 participants among which 32 were residential users and the rest were commercial users. The group of users were provided with IHD along with TOU pricing policy [14]. A post-project survey was conducted for residential and commercial users taking part in the project; 70% of residential and 65% of commercial users stated that their electricity consumption behavior was changed after the installation of the monitoring system which led to less energy consumption.

2.5.3.3 Home Energy Efficiency Trial

In 2004, Country Energy of Australia piloted a project named "Home Energy Efficiency Trial" for a period of 18 months, including 200 households which were provided with IHD and TOU [15]. In general, this kind of monitoring projects is mainly dependent on the time of usage as an indicator and as a metric to measure the amount of the used/consumed energy. This might not be that much accurate as energy should be measured by the Watt-hour unit. Other major studies have also been conducted to determine the effectiveness of smart monitoring system.

The results of the energy savings derived from the aforementioned projects have been compared and categorized according to the type of the system feedback as depicted in Table 2.1.

2.6 OBSERVATIONS AND STATISTICS

The literature suggests that the real-time monitoring yields a high percentage of energy consumption reduction, but when pricing policy is introduced with energy monitoring, extra reduction can be achieved. If the consumers interact with the IHD unit

TABLE 2.1
Energy-Saving Results of the Overviewed Use Cases/Projects

	Reference	Country	Monitoring Period	Energy Saving	Sample Size
Indirect	[18]	USA	2 months	8%	353
feedback	[19]	USA	3 weeks	10%	15
system	[20]	Finland	2 years	3%	525
	[21]	Finland	2.5 years	7%	105
	[22]	Norway	3 years	10%	191
	[23]	Norway	1 year	4%	2000
	[24]	Scandinavia	Unknown	2%–4%	600–1500
Direct	[19]	USA	4 weeks	16%	10
feedback	[25]	USA	11 months	12%	25
system	[26]	UK	4 weeks	9%	80
	[27]	USA	3 weeks	15%	85
	[28]	USA, Canada	Unknown	7%	75
	[29]	UK	5 months	13%	31
	[8]	Japan	2 years	9%	9
	[3]	Canada	2.5 years	6.5%	400
	[4]	Canada	1.5 years	2.7%	200
	[7]	USA	1 years	13%	300
Pay as	[17]	UK	12 months	11%	35
you go	[9]	USA	Unknown	12.8%	100
	[12]	Canada	Unknown	15%	2500
TOU	[17]	USA	Unknown	26%	480
	[13]	Canada	Unknown	7.6%	153
	[14]	USA	Unknown	TBD	61
	[15]	Australia	Unknown	16%	200

effectively, they can reduce their energy consumption averagely 7% (with only IHD method), whereas when using IHD along with prepayment method, energy reduction increases averagely up to 14% [16]. Similar studies as discussed in Section 2.5 have been reviewed, and categorical comparison between them is illustrated in Figure 2.4.

Obviously, energy monitoring is primarily a technique in reserving our planet energy. It allows numerous industries to control their energy consumptions. It eliminates the waste in resources, reduces redundancy in energy usage, and improves existing approaches and procedures in power generation. Essentially, it takes the advantage of statistics and advances in communication technologies. The majority of these overviewed systems consists of the following. First, they consist of a data acquisition system over the cloud obtaining the raw data samples from the aforementioned monitoring devices in order to perform advanced analytics and computations. Second, they utilize wired/wireless communication protocols to transfer the collected/processed data between these monitoring devices and the cloud. Third, they use a data storage unit (or a database) in order to maintain the data for prolonged time periods and use it in future predictions. Finally, they use a GUI (or a web application) in order to provide the remote-friendly access and interact with users.

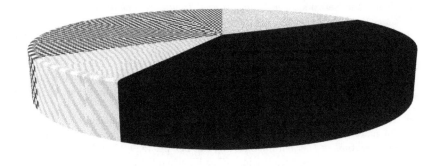

Indirect Feedback **20% Savings**
Direct Feedback (IHD only) **35% Savings**
IHD with Prepay Pricing **15% Savings**
IHD with TOU Pricing **30% Savings**

FIGURE 2.4 Comparison chart between the different monitoring systems.

2.7 CONCLUDING REMARKS

Energy monitoring is essential for the emerging business and varying industries to achieve their goals efficiently. Integrating the cloud/IoT solutions can lead to significant energy savings and improvements. Several research studies and use cases have been conducted in this work, suggesting that the energy-monitoring systems result in more energy consumption reduction. Specifically, the real-time feedback to consumers results in a significant behavior change for energy consumption. TOU pricing policy has been also effective when introduced with IHDs as it yields a higher energy savings. Although the prepay method has its own benefits, but making a hybrid system of both prepaid method and the TOU pricing policy can also yield the highest energy-saving record. It was evident from the literature review that the pilot projects have been very successful. However, most of the projects were implemented only in the USA, Canada, and Scandinavian countries. Therefore, it would be a good starting point to understand the feasibility of implementing such a system worldwide.

ACKNOWLEDGMENTS

We would like to acknowledge all the help and support that have been received thankfully by Eng. Ahmad Rasheed during the course of this work.

REFERENCES

1. F. Al-Turjman, E. Ever, and H. Zahmatkesh, "Small Cells in the Forthcoming 5G/IoT: Traffic Modelling and Deployment Overview", *IEEE Communications Surveys and Tutorials*, 2018. DOI. 10.1109/COMST.2018.2864779.
2. M. Z. Hasan, and F. Al-Turjman, "Analysis of Cross-layer Design of Quality-of-Service Forward Geographic Wireless Sensor Network Routing Strategies in Green Internet of Things", *IEEE Access Journal*, vol. 6, no. 1, pp. 20371–20389, 2018.

3. Hydro One Networks, *The Impact of Real-Time Feedback on Residential Electricity Consumption: The Hydro One Pilot*. Toronto, Ontario, 2006.
4. F. Al-Turjman, and S. Alturjman, "Context-Sensitive Access in Industrial Internet of Things (IIoT) Healthcare Applications", *IEEE Transactions on Industrial Informatics*, vol. 14, no. 6, pp. 2736–2744, 2018.
5. F. Al-Turjman, E. Ever, and H. Zahmatkesh, "Green Femtocells in the IoT Era: Traffic Modelling and Challenges – An Overview", *IEEE Networks Magazine*, vol. 31, no. 6, pp. 48–55, 2017.
6. M. Majidi, B. Mohammadi-Ivatloo, and A. Anvari-Moghaddam, "Optimal Robust Operation of Combined Heat and Power Systems with Demand Response Programs", *Elsevier Applied Thermal Engineering*, vol. 149, pp. 1359–1369, 2019
7. F. Lacey, "Default Service Pricing – The Flaw and the Fix: Current Pricing Practices Allow Utilities to Maintain Market Dominance in Deregulated Markets", *The Electricity Journal*, vol. 32, no. 3, pp. s 4–10, April 2019.
8. J. Truby, "Decarbonizing Bitcoin: Law and Policy Choices for Reducing the Energy Consumption of Blockchain Technologies and Digital Currencies", *Elsevier Energy Research & Social Science*, vol. 44, pp. 399–410, 2018.
9. B. Pruitt, "SRP M-Power: A better way to keep customers in power & save energy", 2019 [online]: https://android-apk.net/app/srp-m-power/1280528250/.
10. J. King, "M-Power, Energy Central, Energy Pulse", 2019 [Online]: https://www.energycentral.com/c/iu/m-power-better-way-keep-customers-power.
11. F. Al-Turjman, and A. Abdulsalam, "Smart-Grid and Solar Energy Harvesting in the IoT Era: An Overview", *Wiley's Concurrency and Computation: Practice and Experience*, 2018. DOI. 10.1002/cpe.4896.
12. "Transmission & distribution", *Woodstock Hydro*, Posted on Sat 21 July 2018 [Online]: http://www.woodstockhydro.com/.
13. Hydro One Networks Inc. Toronto, Ontario, 2019. [Online]: https://www.hydroone.com/.
14. Primen, Inc., "California Road Charge Pilot Program - Technical Advisory Committee (TAC)", California, 2019. [Online]: http://www.dot.ca.gov/road_charge/tac/.
15. CEAU, "Australian Energy Update 2018", Sustainability Victoria, State Government, New South Wales, Australia, 2018. [Online]: https://www.energy.gov.au/publications/australian-energy-update-2018.
16. P. Sabev Varbanov, J. Jaromír Klemeš, and X. Wang, "Process Integration for Energy Saving and Pollution Reduction", *Energy*, vol. 146, pp. 1–178, 2018.
17. I. Laib, A. Hamidat, M. Haddadi, N. Ramzan, and A.G. Olabi "Study and Simulation of the Energy Performances of a Grid-Connected PV System Supplying a Residential House in North of Algeria", *Elsevier Energy,* vol. 152, pp. 445–454, 2018
18. O. Aboelazayem, M. Gadalla, and B. Saha "Valorisation of High Acid Value Waste Cooking Oil into Biodiesel Using Supercritical Methanolysis: Experimental Assessment and Statistical Optimisation on Typical Egyptian Feedstock", *Energy,* vol. 152, pp. 408–420, 2018
19. J. Thakur, and B. Chakraborty, "Impact of Increased Solar Penetration on Bill Savings of Net Metered Residential Consumers in India", *Energy*, vol. 152, pp. 776–786, 2018.
20. S. Alabady, and F. Al-Turjman, "A Novel Approach for Error Detection and Correction for Efficient Energy in Wireless Networks", *Springer Multimedia Tools and Applications*, vol. 78, no. 2, pp. 1345–1373, 2018.
21. F. Al-Turjman, and S. Alturjman, "Confidential Smart-Sensing Framework in the IoT Era", *The Springer Journal of Supercomputing*, vol. 74, no. 10, pp. 5187–5198, 2018.
22. M. Z. Hasan, and F. Al-Turjman, "Analysis of Cross-layer Design of Quality-of-Service Forward Geographic Wireless Sensor Network Routing Strategies in Green Internet of Things", *IEEE Access Journal*, vol. 6, no. 1, pp. 20371–20389, 2018.

23. S. Choudhury, and F. Al-Turjman, "Dominating Set Algorithms for Wireless Sensor Networks Survivability", *IEEE Access Journal*, vol. 6, no. 1, pp. 17527–17532, 2018.
24. L. McClelland, and S. Cook, "Energy Conservation Effects of Continuous In-Home Feedback in All-Electric Homes", *Journal of Environmental Systems*, vol. 9, no. 2, pp. 169–173, 1979.
25. F. Al-Turjman, M. Z. Hasan, and H. Al-Rizzo, "Task Scheduling in Cloud-based Survivability Applications Using Swarm Optimization in IoT", *Transactions on Emerging Telecommunications*, 2018. DOI. 10.1002/ett.3539.
26. K. Bunse, M. Vodicka, P. Schönsleben, M. Brülhart, and F. Ernst, "Integrating Energy Efficiency Performance in Production Management – Gap analysis Between Industrial Needs and Scientific Literature", *Journal of Cleaner Production*, vol. 19, no. 6, pp. 667–679, 2011.
27. Y. Ghadi, M. G. Rasul, and M. Khan. "Potential of Saving Energy Using Advanced Fuzzy Logic Controllers in Smart Buildings in Subtropical Climates in Australia", *Energy Procedia*, vol. 61, pp. 290–293, 2014.
28. S. Alabady, and F. Al-Turjman, "Low Complexity Parity Check Code for Futuristic Wireless Networks Applications", *IEEE Access Journal*, vol. 6, no. 1, pp. 18398–18407, 2018.
29. G. Brandon, and A. Lewis, "Reducing Household Energy Consumption: A Qualitative and Quantitative Field Study", *Journal of Environmental Psychology*, vol. 19, no. 1, pp. 75–85, 1999.

3 Energy Harvesting in IoT-Based Grid

Fadi Al-Turjman
Near East University

Abdulsalam Ahmed Abdulkadir
Middle East Technical University

CONTENTS

3.1   INTRODUCTION

In the 21st century, electricity utilization has changed significantly due to the huge increase in demand. Several uses of the main resources of electricity production surfaced. Hence, the grid of the 20th century becomes inadequate now. The constantly increasing demand for better and effective provision of power resulted in the development of a more robust, effective, and a two-way grid system called "smart grid" [1,2]. Smart grid can be defined as the power generation network that integrates intelligently the utility generators and the end users to efficiently transmit and distribute electricity. It also allows electricity distribution to be sufficient in capacity with good area coverage, safe, economic, reliable, efficient, and sustainable manner [1–3].

Also smart grid is an interconnected network for delivering electricity from producers to end users. It consists of generating stations, and high-voltage transmission lines and distribution lines, with issues resulting from connectivity, being solved in practical scenario using generic approach [4], against machinery or human mistakes.

Solar power is radiant energy received from the sun that is processed using a range of continuously changing technologies; this includes photovoltaic, solar thermal energy, solar heating, solar architecture, molten salt power plants, and artificial photosynthesis [5]. Solar energy is the changing of radiant energy collected from the sun into electric power, either using photovoltaic (PV) directly and concentrated solar panels (CSP) indirectly, or combination. Concentrated solar systems use lenses or mirrors and tracking systems to focus a large area of sunlight into a small beam.

Internet of Things (IoT) is a term mostly given to a set of devices networked together to sense and collect data from everywhere. IoT originated in 1999 when Kevin Ashton first coined the name in the context of supply chain management. Initially, the definition was used to cover lots of ranges that involve applications in healthcare, utilities, transport, etc. But later it changes as technology evolved, and several studies emphasized how computers can sense information without the help of human. When home appliances are connected to a network, they can interact together in cooperation to give the ideal service as a whole and not as an individual entity together working as a single device. This concept referred to as the IoT became very acceptable, useful, and gained application in many areas of human endeavor which do include

- FOOD as smarter food
- TRAFIC as smarter transportation
- ROAD as smarter roads
- HEALTH as smarter healthcare

- BLOOD BANK as smarter blood banks
- GRID as smarter grid.

In the following, we summarize our main contributions and used abbreviations. This chapter overviews potential methods in improving the energy generation process under varying scarcity and uncertainty conditions in the IoT era. It considers optimization aspects of the self-generated power using roof-top solar panels and local energy storage facilities. We discuss the grid evolution, the different types of it, advantages, and what it requires to integrate with the generated solar energy using smart meters, tariff applications, storage devices, and optimization of storage infrastructure. Moreover, Table 3.1 shows the list of abbreviations in this chapter.

TABLE 3.1
Abbreviations Used in the Article

Abbreviation	Description
TWh	Terawatt hour
GHG	Greenhouse gas
PV	Photovoltaic
CSP	Concentrated solar panels
IoT	Internet of Things
KW	Kilowatt
MW	Megawatt
FLISR	Fault location isolation and service restoration
FACTS	Flexible alternating current transmission system
RES	Renewable energy sources
AMI	Advance metering infrastructure
AMR	Automatic meter reading
WAN	Wide area network
HAN	Home area network
3G	Third generations
GSM	Global System for Mobile
NIST	National Institute for Standards and Technology
ToU	Time of use
CAES	Compressed air energy storage
NaS	Sodium sulfur
EVs	Electric vehicles
HEVs	Hybrid electric vehicles
FHEV	Full hybrid electric vehicle
MHEV	Mild hybrid electric vehicle
BEV	Battery electric vehicle
FCEV	Fuel cell electric vehicle
PHEV	Plug-in hybrid electric vehicle
MHV	Microhybrid vehicle
V2G	Vehicle-to-grid

3.2   LITERATURE REVIEW

The first power grid was built in 1895 with the main aim of supplying electricity. Presently, there are 9200 power grids all over the world producing and supplying consumers with about 1 million MW of electricity on daily basis [1]. The grid starts from just a conventional one to a microgrid and the smart grid, and the conventional grid was developed for convention power generation from fossil fuel generating plants, nuclear plants, hydro power plant with significantly high power loss due to long transmission lines, while the others were as a result of advancement in electricity production from solar and wind sources with optimum aim of effective power transmission. The microgrid is quite small which focuses more on local electrification and can stand outside the conventional grid or integrate with it. Its transmission of electricity is across short distances; therefore, loss in energy encounter during transmission in the conventional grid is much reduced with capacity range of 3–10 MW.

In Nigeria, electricity generation started as early as 1896, with the installation of its first power plant of just 60 kW generating capacity that was more than the energy need then, in Lagos in 1898 [6,7]. Then with series of expansion followed by neglect of infrastructural development, reported vandalizations and of course corruption till today the energy need for the continuously increasing population was never satisfied even after huge spending by different administrations. Presently, officials of the Administration of President Muhammadu Buhari reported electricity availability of 7000 MW that is quite poor for approximately 186 million citizens (as of 2016), but available reports indicated just above half of that value (3941 MW) is available from the 23 power plants connected to the National Grid as at May 2015.

3.2.1   SMART GRID

Smart grid literally means upgrading the traditional grid from one-way source to two-way source electricity generation, which is as well digitalized, modernizing with well-detailed automation for the 21st-century electricity transmission and distribution by simply adding censors and software or programs to the conventional grid that enriches providers of electricity and end users with lots of information to work with in effective transmission and distribution of the electricity and certainly effective, low-cost utilization of the electricity by providing more knowledge when electricity is cheap to use and expensive.

Smart grid systems can be described as a combination of distributed and centralized generator which are utilized to control the low and high level voltage distribution through automation systems of the different users more secured and reliable energy to these users for their respective uses [1]. Smart grid was as well referred by [8], as a system that recently revolved using available heterogeneous networks (HetNets) to control our huge electricity needs in smart, economical, and sustainable way. This create competition in energy market allowing different sources, such as solar and wind energy sources including better monitoring of transmission facilities over regions, and most importantly, it allows the possibility of end users to act as a service provider [9].

3.3 ADVANCEMENTS IN THE SMART GRID

It is the integration of traditional electric grid with communications and information systems to better monitor and control the flow and consumption of energy, save energy, reduce cost, and increase reliability of electricity generation and transmission, thus making it modern electric grid which makes it possible to connect electricity generated from non-conventional sources.

In a smart grid system, sensors are deployed linearly, and programs such as Fault location isolation and service restoration (FLISR), a distribution automation tool, help in optimization plan for switching action as well as restoring supply, also flexible alternating current transmission system (FACTS), that help in controllability enhancement including increase in power capability will send notification if a grid shuts down the service provider will be notified with all details regarding such [10]. It was reported that the integration of renewable energy sources (RES) helped significantly in reducing system losses; increased the reliability, efficiency, and security of smart grid; and made its infrastructure fully advanced in sensing and communication computing abilities. Hence, it can be said that smart grid reduces greenhouse gas (GHG) emission by allowing the integration of renewable sources to the grid, and thus, the dependence on fossil fuel decreased. The smart grid uses the synchrophasor technology to prevent constraint, system collapse, natural incidents, and equipment failures which are the general conditions that cause energy disturbance, and actually, it can be thus said that this system can understand the situation and know in advance the challenges, thereby resolving crises before it even occurs [1,2].

The enhanced benefit of smart grid is in smart metering that provides a two-way advantage to both providers and end users. The smart meter provides the service providers with detailed hourly utilization of energy by end users and uses the collected data in billing the home owner as well as balancing the consumption of power by reducing price for home owner at certain hours of the day to control peak energy consumption. Mainly smart meter consists of the following components: analog and digital ports, serial communication, volatile memory, non-volatile memory, real-time clock, and microcontrollers.

3.3.1 BENEFITS OF THE SMART GRID TO ENERGY SECTOR

The benefits of the 21st-century grid that is referred to as smart grid are numerous. It is the most important smart meter in advance metering infrastructure (AMI), and some added incentives provided by upgrading to smart grid are listed as follows:

- Providing customers the opportunity to have an effective financial plan on electricity usage and reliable electricity
- Providing efficient, robust, and qualitative electricity, and efficient renewable power generation
- Enhancing energy sources mix
- Creating interface for smart devices in home automation
- Reducing carbon footprint

- Creating enabling environment for electric vehicles and lots of job opportunities
- Anticipating faults or problems
- Self-healing response ability
- Providing alternative for storage and security of infrastructure.

3.4 METERING TECHNOLOGIES

The meters used initially in electricity transmission, distribution, and supply were actually to allow means of knowing the amount of consumed electricity for billing purposes only. This type of meters is referred to as traditional meters, and the technologies used by such meters are called electromechanical technology. The need to provide more efficient and reliable electricity resulted in the development of smart meters, such as automatic meter reading (AMR) meters, and later advanced metering infrastructure (AMI) meters. The different types of meters from traditional AMR to AMI meters are shown in Figure 3.1. The AMR meters uses the electromechanical plus electronic technology that allows providers to read electricity consumption remotely using fixed network abilities such as hand-held/walk-by or drive-by devices. It still did not provide the much needed two-way advantage that allows consumers active participation, and this brought the advent of the AMI using the newest technology in metering referred to as hybrid technology.

3.4.1 GAIN OF SMART METERS VS. THE CONVENTIONAL METER

- The conventional meter can only provide usage data to service provider, in most cases, collected manually for monthly billing.
- On the other hand, smart meter provides what is referred to as two-way information that helps the utility provider and the end user.
- It informs both parties of any power outage and possible theft.

FIGURE 3.1 Different types of meters.

- It also improves power quality, gives more efficient and secure power delivery, enhances electricity reliability, and allows for prepayment options and more accurate billing.
- In short it brings the end of estimated bills and over paying/underpaying.

3.5 COMMUNICATIONS USED IN SMART GRID

The smart meter is connected in two ways using wide area network (WAN) and home area network (HAN). WAN is used to connect smart mater, supplier, and utility server, whereas HAN is used for connecting smart meter with home appliances. The different technologies used by WAN for communications are fiber optics and 3G/GSM, whereas technologies used by HAN are Bluetooth, Zigbee, and wireless Ethernet or wired Ethernet.

3.5.1 ZIGBEE

In wireless communication technologies, National Institute for Standards and Technology (NIST) confirmed Zigbee Smart Energy Profile as the most applicable communication infrastructure for smart grid network due to its advantages such as low cost, comparatively low-power consumption, less complexity, and fast data rate transfer. Zigbee is used in smart grid for automatics meter reading, energy monitoring, and home automation. It has a bandwidth of 2.4 GHz plus 16 channels, and every channel uses 5 MHz bandwidth and a maximum output power of 0 dBM including transmission range between 1 and 100 m with a data rate of 250 kb/s [1].

3.5.2 WIRELESS MESH

A wireless mesh network is a combination of nodes that are joined in groups and working as a self-reliant router. Self-healing property of these nodes helps communication signal find a route through active nodes. Infrastructures of mesh network are decentralized because each node sends information to the next node. Wireless mesh is used in small business operation and remote areas for affordable connections [1,10,11].

3.5.3 GSM

GSM is Global System for Mobile communication used to transfer data and voice services in communication technology. It is a cellular technology that connects mobile phone with the cellular network [1].

3.5.4 CELLULAR NETWORK

Cellular networks can be another excellent option for communication between far nodes for utility purpose. They are used to build a dedicated path for communication infrastructure to enable smart meter deployment over a WAN [1].

3.6 BILLING METHODS

Majorly, there are three different types of billing in the smart grid system: net metering, feed-in tariff, or time of use (ToU). These are discussed next.

3.6.1 NET METERING

This is a mean of measuring the amount of excess energy a homeowner was able to produce to the smart grid which is returned back to the homeowner at the same tariff rate during night or when there is less sunlight for him to produce enough energy.

3.6.2 FEED-IN TARIFF

In this situation, a home owner is paid a rate which is usually calculated using the ToU. Should he need energy back, it will be as well sold to him at similar rate.

3.6.3 TIME OF USE

This is a billing plan in which the electricity consumption is calculated based on the real time in which such energy is utilized. The pricing of electricity usage is significantly low when used at time that the demand for power supply is low and high when used at time or period of high power demand. When a homeowner shifts his large electricity consumption to period of low demand, he capitalizes on paying low energy bill [12].

The time of the day when energy demand is high is called on-peak, and time of the day when the demand is low is referred to as off-peak, while in-between these periods is the mid-peak. There are variations in timing for different seasons: for summer, peak time is 3 pm–8 pm and mid-peak time is 6 am–3 pm, whereas for winter, peak time is 6 am–10 am and mid-peak time is 10 am–5 pm. The best plan is usually to move more electricity usage to off-peak which is more preferable between 11.30 am and 2.30 am [13].

Table 3.2 shows how price of electricity can be reduced by moving most electricity consumption to off-peak periods by computing ToU pricing for 1000 kWh used per

TABLE 3.2
Portland General Electric ToU Price Compared with Standard Rate

Change of Usage Period of Electricity	On-Peak (kWh)	Mid-Peak (kWh)	Off-Peak (kWh)	Time of Use (ToU)	Standard Rate
To continuously use	206	408	386	$68.06	$68.50
When 10% of on-peak is shifted to off-peak hours	185	408	407	$66.21	$68.50
When 25% of on-peak is shifted to off-peak hours	154	408	438	$63.48	$68.50
When 35% of on-peak is shifted to off-peak hours	134	408	458	$61.73	$68.50

month at different periods with a table showing variation in the seasonal pricing periods.

Pricing for each time of the day is as follows:

- On-peak 13.197 cent per kWh
- Mid-peak 7.572 cent per kWh
- Off-peak 4.399 cent per kWh

This will bring to an end the usual fixed monthly charge plan where electricity will be charged based on its usage.

3.7 SOLAR ENERGY

The energy from the sun is certainly enormous and will of course remain forever, the process of collecting the energy for use as electricity varies, and it is such type of generated energy that is referred to as solar energy or solar electricity. The major ways of utilizing sun radiations to generate electricity are discussed next.

3.7.1 SOLAR PANELS (PHOTOVOLTAIC MODULES, PV)

This is the most common method of generating electricity from the sun. It is mainly the use of materials that are semiconductors that easily give out electrons when it absorbs heat from the sun, to produce what is called a solar panel. This has been used in power houses, industries, and the largest PV generating plant (290 MW plant) in Arizona [14].

3.7.2 SOLAR THERMAL (CONCENTRATED SOLAR POWER, CSP)

This technology used to generate electricity from the sun is not very popular and will produce far more electricity than the PV; thus, it cannot be utilized for residential purposes but for large-scale purposes, i.e., in utility plants. The largest facility using this technology to generate electricity is the 377 MW plant in a desert in California

The concentrated solar power (CSP) is categorized into different known technologies based on the collection of energy from the sun, which are discussed next.

3.7.2.1 Dish Engine Technology

In this method, power is produced from the use of dish-shaped parabolic mirrors where gas is heated in a chamber with collected energy from the sun and the heated gas drives a piston of a generator, thereby producing electricity. For more efficiency, the dish is mostly attached to a tracking system that maximizes sun radiation [13].

3.7.2.2 Parabolic Trough

This method uses trough-shaped mirrors which are long enough to reflect solar energy into a tube filled with liquid which collects heat as it focuses on the sun, thereby heating the liquid in the tube connected to heat exchange system where water is heated to steam that is used in a steam turbine generator to generate electricity.

This operates in a continuous recycling process: as the heated fluid transfers its heat, the steam cools and condenses; then, the same process is continuously repeated over and over. Heated fluid can also be stored for a long period and then be reused when the sun is down [9,13].

3.7.2.3 Tower Focal Point-Concentrated Solar Power

In the focal point CSP technology, a large number of flat computer-controlled mirrors constantly move to get maximum sun reflection all through the day onto a tower that has a collecting tank on it. Molten salt moves in and out of the tank and as a result are heated to a very high temperature of 537.8°C (1000°F); the heated fluid is sent to a steam boiler where it uses the inherent heat in the fluid to propel a steam turbine to produce electricity [13].

3.8 STORAGE FACILITIES

The usage of energy from the sun gained significant importance with improved storage facilities such as batteries (e.g., sodium sulfur, lithium ion, lead–acid, metal–air), compressed air energy storage (CAES), pumped hydro, flywheels, fuel cells, and super capacitors [15].

3.8.1 SODIUM SULFUR (NaS) BATTERIES

This is manufactured from a combination of two salts: sodium and sulfur. Excellent energy density, very good charge and discharge efficiency, and low cost make it advantageous for use as storage facility in smart grid.

Sodium sulfur batteries are also used as utility-scaled energy storage in Japan, with hundreds of MW as shown in Figure 3.2. NaS batteries were first hosted as demonstration project in 1992 which become available commercially in 2002 [16].

3.8.2 FLYWHEEL STORAGE DEVICE

This revolution in technology of storage facility is regarded as the most used for smart grid storage infrastructure at both transmission and distribution stages. This could

FIGURE 3.2 Large sodium sulfur (NaS) batteries [16].

FIGURE 3.3 Rotating flywheel storing energy mode [17].

be a result of its rapid response time in discharging stored electricity, high efficiency, long life and the quite insignificant maintenance. It is cylindrically shaped, having large rotor contained in a vacuum. Flywheel storage device stores electricity in the form of rotational energy, which can also be referred to as spinning mass.

It can store electricity for a smart grid by drawing energy from it as the rotor rotates at a very high speed, as shown in Figure 3.3. To discharge the electricity, it slows down by switching to generation mode, thus returning electricity back to the grid for onward distribution [17].

3.9 OPTIMIZATION OF STORAGE DEVICES

It has been emphasized that it is very importance to stabilize the energy in smart grid connection due to the intermittent nature of generated energy from the renewable sources. The best form of storage is eminent, should there be urgent need to balance the grid's energy gap. For such optimization, electricity storage with rapid response time is very important. And the way in which electricity is stored (e.g., in form of electrical, mechanical, and/or electrochemical), as well as the location of the storage unit is of utmost importance also. The storage device has different applicable usage in the grid stabilization; for instance, sodium sulfur batteries are very reliable and are used in centralized energy storage systems, pumped hydro storage and CAES systems are used in smart grid for load leveling applications and transmission services, while flywheel batteries are used for power quality applications. The battery energy storage system helps significantly in stabilizing smart grid with larger quantity of electricity from renewable source by storing the energy when in excess and returning it to the grid when required [18].

Batteries are classified into mobile and stationary batteries; the stationary batteries take surplus electricity from the grid and store it in the form of electrochemical batteries and give back electricity to the grid when there is demand [18].

On the other hand, mobile batteries are used in the form of electric vehicle (EV) batteries which has an added advantage in reducing CO_2 and storing electricity when energy is cheap and returning it when expensive. Presently, there are two varieties: the

EVs and hybrid electric vehicles (HEVs) which can further be divided into six types due to the energy demand as well as power of the batteries [19]. However, charging the electric vehicle could be a problem on managing grid peak, but the advent of decentralization using multiagent systems will provide solution to such problems [20].

EVs are classified into two classes, whereas HEVs are classified into four classes [18]. EVs classes are: (1) battery electric vehicles (BEV), and (2) fuel cell electric vehicles (FCEV). As for the HEVs, they are classified into (1) plug-in hybrid electric vehicle (PHEV), (2) microhybrid vehicle (MHV), (3) full hybrid electric vehicle (FHEV), and (4) mild hybrid electric vehicle (MHEV). Out of all, only two, BEVs and PHEVs, are suitable for vehicle-to-grid (V2G) operation. V2G technology enables the transmission of power between vehicles and the power grid by utilizing the on-board batteries of BEVs and PHEVs as devices for storing energy in the smart grid systems [18].

High optimization can be achieved by optimizing the feet charging load in a power grid with reference to the following factors: (1) generation capacity, (2) load characteristics, (3) power quality, and (4) grid reliability. Claudio et al. [21] presented and considered some strategies in energy storage which are given as A-posteriori optimal strategy algorithm (AOS), dynamic programming with Markovian request (DPM) algorithm, single-threshold (ST) algorithm, and dynamic programming with independent request (DPI) algorithm. The strategies need less information on end-user power statistics request but provide reduction in price. The following were noted that AOS is certainly not realistic in practical situations, DPM accuracy can be increased by adding other parameters but its optimality is not guaranteed any more, but after simulation, it highlighted the following:

- ST algorithm in realistic power and battery demand mode is not optimal
- DPM did achieve noticeable performance gain against DPI
- DPM could be near AOS upper bound if realistic power is requested

Therefore, declared DPM as best strategy in energy storage [21]. Summarily stored energy can as well be referred to as produced energy [22].

3.10 CONNECTING RENEWABLE (SOLAR) ENERGY TO THE SMART GRID

As solar technology continues to improve in efficiency and become more popular, the cost is expected to decrease significantly; thus, the payback period of a PV system that stands at an average of 20 years could drop to 10 years. This means even without government incentives it will be cheap to own PV system which could be cheaper when the system is connected to grid as batteries increase the price due to the fact that it needs replacement. Major equipment, referred to as balance of system, to integrate with a state utility smart grid or a national smart grid are (1) power conditioning equipment, (2) safety equipment, and (3) instrumentation and meters. Power conditioning equipments practically change DC to AC as electronics devices uses electricity in AC mode. The major power conditioning equipments are

- Constant DC power to oscillating AC power conversion
- Frequency of the AC cycles should be 60 cycles per second
- Voltage consistency is the range allowed for output voltage to fluctuate
- Quality of the AC sine curve, regardless of the fact that AC wave shape is jagged or smooth

Some electric appliances can operate regardless of the electricity quality, whereas others need stabilizers to operate. Inverters are devices required to stabilize the intermittent electricity so that it harmonizes with the requirements of the load in the grid. Safety equipments are devices that provide protection to stand alone or home owner's system and on-grid integrated electricity generated through solar or wind source from being destroyed or endangering people during natural occurrences such as storm, lightening, power surges, or fault that can result from faulty equipment. Instrumentation and meters are equipments that provide home owner with the ability to view and manage their electricity in form of solar energy, the battery voltage of his system, the quantity of electricity one is consuming, and also the strength of the battery in term of charge/discharge.

3.11   GRID TOPOLOGY

Based on the environment as well as the geography of the area a grid is covering, it is divided into (1) radial distribution, (2) mesh distribution, and (3) looped or parallel flow distribution.

3.11.1   RADIAL DISTRIBUTION

This is referred to as the cheapest to install and simplest form of topology in grid transmission and distribution which is used generally in scattered populated locality. It is shaped like a star, showing power distribution from one major supply to the end users. In this topology, a single fault can result in a total black out. Example is the traditional grid system, as shown in Figure 3.4.

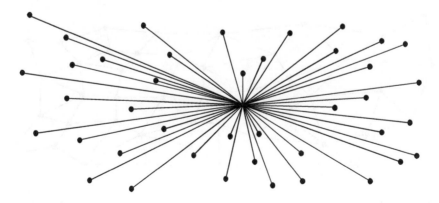

FIGURE 3.4   Radial topology.

3.11.2   Meshed Distribution

This is a better topology with more efficient means of transmitting and distributing power. The infrastructure devices are interconnected to allow for cooperation between power source. This makes power rerouting possible, in an event of a failure between the source and the end user. It is mainly applied in a high loaded area that is quite congested. Figure 3.5 shows the possibility, for example, of rerouting the power distribution in a congested locality.

3.11.3   Looped Distribution

This is the most expensive topology mostly found in European countries including Northern American town and cities. It is more reliable as it provides the ability of rerouting power in case of fault in a power line before repairs, thereby allowing continues service, as represented in Figure 3.6. From its name, it connects two power sources to an area by reconnecting it to the original power source; as a result, an alternative power source is always available.

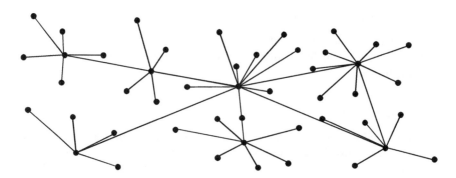

FIGURE 3.5   Meshed topology.

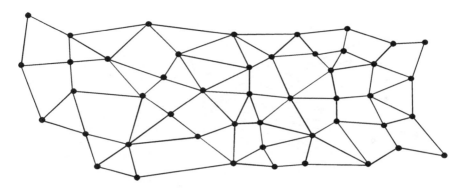

FIGURE 3.6   Looped topology.

3.12 CONCLUSION

Energy provision of any country is universally accepted to be the yard stick in measuring her development, either under developed, developing, or developed. It is therefore necessary for Nigeria to take the right steps towards bettering the energy availability to her citizens. Presently, most Nigerians produce their own electricity which is majorly from fuel generator with few producing from roof top solar PV. It was reported that just 37 universities in the country use over a thousand generators. Nigeria is rated globally as the highest importer of generators. We believe that if homeowners can get credit by their local power companies for the electricity produced at their homes through "net metering" or "feed-in tariff" programs, it will encourage more homeowners to participate and eventually increase greatly the energy generation and significantly reduce the greenhouse gas emission.

The IoT smart grid promises to break these two entry barriers by offering low-cost flexible solutions and using new communication solutions coming from the Internet world, thereby providing a more efficient energy production and utilization which will significantly provide the most needed security for the power sector infrastructure against external attack.

REFERENCES

1. M. I. U. Khan and M. Riaz, "Various Types of Smart Grid Techniques: A Review," *International Journal of Multidisciplinary Sciences and Engineering*, vol. 7, no. 8, p. 7, 2016.
2. A. Faizan, "Smart Grid Technology and Applications Smart Grid Components," *Electrical Academia*, 12-Nov-2017.
3. "Building the smart grid," *The Economist*, 04-Jun-2009.
4. F. M. Al-Turjman, H. S. Hassanein, and M. Ibnkahla, "Quantifying Connectivity in Wireless Sensor Networks with Grid-Based Deployments," *Journal of Network and Computer Applications*, vol. 36, no. 1, pp. 368–377, 2013.
5. E. Camacho, T. Samad, M. Garcia-Sanz, and I. Hiskens. Control for renewable energy and smart grids. *International Journal of Impact Control Technology,* vol. 6, 19–25, 2011.
6. U. P. Onochie, H. O. Egware, and T. O. Eyakwanor, "The Nigeria Electric Power sector (Opportunities and Challenges)," *Journal of Multidisciplinary Engineering Science and Technology*, vol. 2, no. 4, pp. 494–502, 2015.
7. F. Al-Turjman M. Abujubbeh, "IoT-enabled Smart Grid via SM: An Overview," Elsevier Future Generation Computer Systems, vol. 96, no. 1, pp. 579–590, 2019.
8. F. Al-Turjman, "Mobile Couriers' Selection for the Smart-Grid in Smart-Cities' Pervasive Sensing," *Future Generation Computer Systems*, vol. 82, no. 1, pp. 327–341, 2017.
9. ASEA Brown Boveri, "Industrial smart grid," 3AFA 701889, 2011.
10. United States Department of Energy, "Balance-of-system equipment required for renewable energy systems | Department of Energy," [Online]. Available: https://www.energy.gov/energysaver/balance-system-equipment-required-renewable-energy-systems. [Accessed: 17-May-2018].
11. O. Patrick, O. Tolulolope, and O. Sunny, "Smart Grid Technology and Its Possible Applications to the Nigeria 330 kV Power System," *Smart Grid Renew. Energy*, vol. 4, no. 05, p. 391, 2013.

12. M. M. Eissa, "Introductory Chapter: Demand Response Incentive Program (DRIP) with Advanced Metering and ECHONET," in *Smart Metering Technology and Services-Inspirations for Energy Utilities*, InTech, 2016.

13. Portland General Electric, "Time of Use Pricing - Time of Use PGE," [Online]. Available: https://www.portlandgeneral.com/residential/power-choices/time-of-use/time-of-use-pricing. [Accessed: 17-May-2018].

14. Luis Rejano Flores, "Difference between solar thermal and photovoltaic power," *The Energy of Change*, 11-Aug-2014. [Online]. Available: http://www.theenergyofchange.com/difference-between-solar-thermal-and-photovoltaic-power. [Accessed: 17-May-2018].

15. "How energy storage works," *Union of Concerned Scientists*. [Online]. Available: https://www.ucsusa.org/clean-energy/how-energy-storage-works. [Accessed: 17-May-2018].

16. A. Price, S. Bartley, S. Male, and G. Cooley, "A Novel Approach to Utility-Scale Energy Storage," *Power Engineering Journal*, vol. 13, no. 3, pp. 122–129, 1999.

17. W. Saad, Z. Han, H. V. Poor, and T. Basar, "Game-theoretic Methods for the Smart Grid: An Overview of Microgrid Systems, Demand-Side Management, and Smart Grid Communications," *IEEE Signal Processing Magazine*, vol. 29, no. 5, pp. 86–105, 2012.

18. C. Kang and W. Jia, "Transition of tariff structure and distribution pricing in China," in *2011 IEEE Power and Energy Society General Meeting*, 2011, pp. 1–5.

19. Y. S. Wong, L. L. Lai, S. Gao, and K. T. Chau, "Stationary and mobile battery energy storage systems for smart grids," in *2011 4th International Conference on Electric Utility Deregulation and Restructuring and Power Technologies (DRPT)*, 2011, pp. 1–6.

20. M. Morte, "E-mobility and multiagent systems in smart grid," in *Electric Power Engineering (EPE), 2016 17th International Scientific Conference on*, 2016, pp. 1–4.

21. P. Vijayapriya, G. Bapna, and D. D. P. Kothari, "Smart Tariff for Smart Meters in Smart Grid," p. 6, 2010.

22. J. Das and S. Ashok, "Peak load management with wheeling in a combined heat and power unit under availability based tariff," in *Advances in Computing and Communications (ICACC), 2014 Fourth International Conference on*, 2014, pp. 343–346.

4 Grid Energy Scenario and Storage Systems for the Vehicle-to-Grid Technology
An Overview

Fadi Al-Turjman
Near East University

B. D. Deebak
Vellore Institute of Technology

CONTENTS

4.1   INTRODUCTION

The world population is rapidly growing that increases the usage of energy consumption and greenhouse gas (GHG) emission year by year. Improving the amount of installation of renewable energy sources for the distribution and production of energy resources may extend the use of energy efficiency. As there is no means to determine the effective storage of excessive energy source generated by the wind blows and the sun shines, the energy production is highly relied on for non-renewable energies. A proposed strategy known as energy wastage integrates the vehicle-to-grid (V2G) technology to convert in the form of the energy system. It is nowadays providing energy storage capacities for electric vehicles (EVs). The EVs are much adopted for precisely 12.1% of transportation system. The specific factors such as load balancing and frequency regulation [1] are beneficial for the EV system and the vehicle owner, which potentially be able to generate a profit on electricity over due course of a period. Also, this usage ensures that it can provide better service to an electric grid.

However, several strategies should be considered to integrate the V2G charging that makes the energy grid system to promote smart intelligent systems [2–7]. The utilization of V2G technologies and its full-scale deployment integrates a smart system to communicate over networks. This integration tries to provide a feasible business model in turn to ease the complexity of the energy distribution network. The EV's deployment plays a significant role to achieve mass production service that potentially enhances the technological development of the V2G system. V2G is not aimed to promote the individual's business rather promoting the collaboration among various EV owners. Therefore, it is said to be a subdivision of electric grids and fleet systems. The effect of integration provides specific features to enhance high storage capacity and low GHG emission. According to Paris Agreement [8], the contribution of UNFCCC 2015a depreciates the GHG emission around 2°C global temperature. Moreover, an International Energy Agency (IEA) 2°C Scenario (2DS) trajectory sets a criterion that reduces the GHG emission from 33 G_tCO_2 to 15 G_tCO_2 approximately in 2050, which was roughly around 45% in 2013.

Some IEA predictions and UNFCCC planning strategies of EV are as follows [9]:

1. Around one billion EV comprising of above 40% of LDV stock, which is specifically set with the criterion of 2DS trajectory.
2. Above 400 million electric two-wheeler vehicles are aimed to produce in 2030.
3. By 2050, all the cars will be equipped with EV system.
4. Sixteen governments are agreed to be a member of EVI
5. In the mid of 2014 and 2015, the new EV enrolment system has increased by 70%, i.e., more than 500K sold worldwide.
6. The annual sale list has increased more than 75% in 2015 in countries such as Germany, France, Korea, Sweden, Norway, UK, and India.

7. In 2008, the cost of plug-in hybrid electric vehicle (PHEV) batteries was USD 1000/kWh, which was declined to USD 268/kWh in 2015. In 2022, the target cost is expected to be USD 125/kWh.
8. In 2008, the PHEV battery density was 60 Wh/L, which was amplified to 295 Wh/L in 2015. In 2022, the target density is set to be 400 Wh/L.

Renewable energies such as wind and solar energy have become an important role for the zero GHG emission of production energy [10]. The supply of energy grid types is more flexible rather fossil energy [11]. Various renewable energies are available to prefer because they are not stable energies. However, their construction can be of any size and everywhere. The supplementary such as smart grid, micro-grid, and V2G can be expanded to each other including renewable energy conversion. The DC energy supplies and EV's energy storage systems play an important role to fulfill the energy exchange in the grid system. The EV's energy can significantly be saved in the mid of 23:00 to 5:00 which strongly decreases the grid energy. Hence, the energy level reduction is not economical in a power plant. On the other hand, the ICE efficiency of fossil fuel provides the best condition with the latest technology around 18%–20%. However, the efficiency of a fossil power plant and CHP energy is reported as 38%–40% and 60%–75%, respectively [12].

The EV's benefit is not concerned about zero GHG emissions. Therefore, EVs are operated with 38%–75% of electric energy generated that depends on the conditional energy generation not only to enhance energy efficiency but also to reduce GHG emission. This technology is completely grounded on the use of bidirectional energy distribution systems that are easy to reach in terms of stability, energy reliability, and peak reduction. EV's batteries, quantities, charging time, and power system topologies are specifically employed to charge the energy storage system that is sensible in the use of DC lines in the grid. This situation conveys that the grid energy is critical to stabilizing the energy storage such as peak and downtime in charging station and temporary energy storage system in EVs [13,14]. Thus, this chapter will appraise various energy scenarios, energy storage systems, V2G infrastructure models, and their market influential factors.

4.2 ENERGY SCENARIOS IN V2G TECHNOLOGY

Each energy system has a specific scenario to plan. GHG emission and energy efficiency are considered as the targets for each energy scenario across the globe where the recent scenario known as V2G technology has owned its location dependency, and thus, it has a different set of requirements and energy stocks that exist all across the world. IEA has created a roadmap for 2050. Table 4.1 shows selling numbers of EVs and PHEVs in 2050 by IEA.

It is observed that EVs have lower GHG emissions rather PHEVs emissions. According to the IEA roadmap, Europe and North America have less PHEV emission than China and India. This is meant that the consumption of fossil energy in the transportation sector is lesser than in China and India. For instance, the

TABLE 4.1

Selling Number of EVs and PHEVs in 2050 by IEA

Location	EV	PHEV
India	8600K	9600K
Pacific	2400K	1300K
China	9400K	11,400K
Europe	6400K	3100K
North America	8800K	3800K

government of China tries to decrease GHG emissions concerning other countries. But, the other countries except China have updated their energy system in the form of cybersecurity, smart grid, and micro-grid [15]. As of today, the energy suppliers provide the energy in several forms such as oil, gas (LNG, CNG, and LPG), or electricity. Therefore, the usage of energy methods plays a key role to observe the factor of energy consumption. The percentage of energy consumption for the various factors could be available as follows: electricity 30%, transport 30%, and heat 40%.

In 2050, the research project of Coherent Energy and Environmental System Analysis (CEESA) ensures that all the scenario will be supplied with renewable energy source [16]. Supply control and energy demand define the primary parameters for the energy quality that efficiently analyzes the energy efficiency and reduces the emission of GHG accurately. These are precisely controlled to facilitate the renewable energy network to grid and EVs. Moreover, this precision factor supplies various data related to energy to supply/fulfill the demand of various infrastructure grid system. Above all, it may have some additional classifier techniques to generate a predictable data pattern. A platform such as grid infrastructure will try to obtain a reasonable relation among economical condition and technical benefits of the investor. Thus, V2G technology is preferred to be more economical for various energy storage systems, which have size and quality as the influence factors.

The above factors are completely controllable but can't be predictable. To predict the data precisely, the management team can intellectually raise the schedule timing over demand. Some scenarios may be reviewed one after another to issue a special condition and location, and in addition, it is helpful to change the main scenario and generate a different resulted outcome. According to the scenario outcome, the grid infrastructure design should be more flexible, manageable, and deployable, but it will be more comprehensive and complex. The grid design has a specific algorithm that provides a connection to control the smart infrastructure over design protocols [17].

4.2.1 STORAGE SYSTEM IN V2G TECHNOLOGY

This section will discuss various energy storage types and charging systems. The former briefly explains the use of a battery in EVs, where a mechanical strike, temperature stability, storage controller, and battery longevity will be explained pointedly. The latter one concisely gives the details of charging systems.

4.2.1.1 Preferable Batteries for EVs

Each situation demands its energy storage types. As an instance, the hydro-power plant uses water pumping to save energy sources. This storage parameter categorizes the advanced parameter such as technical aspects involving power rating, density, energy response time, efficiency, charge, discharge, and lifetime. So, EV must be solid to categorize the types in terms of NiCd, Li-ion, ZnBr, and NaS. The important properties of EV's properties are shown in Table 4.2 [18]. However, the cost of each battery is varied to signify the issue related to technical aspects. Li-ion batteries have the alloys such as Fe and Mn to improve the device performance, i.e., storage capacity. In EV, the batteries namely $LiFePO_4$, $LiCoO_2$, and $LiMn_2O_4$ cover a distance between 250 and 350 miles that produce the types in terms of anode and cathode. Table 4.3 shows the Li batteries and related specifications [19].

Depending on the alloy usage, the battery performance can be varied to grade the quality and EV manufacturers. They have some key parameters such as voltage, current, temperature, and mechanical strike to check the test results of the batteries.

4.2.1.2 Mechanical Strike

This issue is completely indispensable to verify whether the battery works at normal or abnormal conditions. At a normal state, the EV has an excessive battery use to influence the effect of mechanical strike, whereby it has an unreliable behavior

TABLE 4.2
Storage Parameters and Specifications

Parameters	NiCd	NaS	ZnBr	Li-Ion
Power rating (MW)	0–40	0.05–8	0.05–2	0–0.1
Discharge timing	S–h	S–h	S–10h	Min–h
Energy density (M/W)	15–8	15–300	65	200–400
Power density (Wh/I)	75–700	120–160	1–25	1300–10,000
Response time	<S	<S	S	<S
Efficiency (%)	60–80	70–85	65–75	65–75
Lifetime cycles	1500–3000	2500–4500	1000–3650	600–1200

TABLE 4.3
Li-Ion Batteries and Specifications

Parameters	Energy Density	Life Cycle	Safety
$C/LiCoO_2$	110–190	500–1000	Poor
$C/LiMn_2O_2$	100–120	1000	Safer
$C/LiFePO_4$	90–115	>3000	Very safe
$LTO/LiCoO_2$	70–75	>4000	Extremely safe
$LTO/LiFePO4$	−70	>4000	Extremely safe

in case of any accident. Therefore, the test standards such as SAE-j2464, T4, and FMVSS-305 are conformed to absorb the battery test [20]. The battery has 80% system on chip (SOC) ratio to address the battery lifetime.

In 2013, JA 829J (Japan Airline Flight) Boeing 787 integrated APU (auxiliary power unit) to show the thermal runaway, i.e., 52,000 flight hours. This flight battery type was lithium cobalt oxide ($LiCoO_2$) until grounded. According to the US-NTSB report, the APU battery had T5 and T6 mechanical absorbance resulting in a fire breakout.

4.2.1.3 Temperature-Stability

This parameter is considered to differentiate the battery behavior, i.e., concerning the temperature. It cannot perfectly work from −120°C to 130°C to address the issue of runaway threat, i.e., abnormal situation. Therefore, the EV batteries should protect against the extreme temperatures, i.e., heating and cooling system, and mechanical-strike effect [21].

4.2.1.4 Storage Control Systems

In the electric grid, grid integration plays a vital role to achieve better power efficiency. In general, the storage system demands more controllers than the other parts of the system grids. Therefore, the storage system is so critical to ensure whether it is healthier or not. In the case of unlimited grid storage, the EV batteries have a grid storage part to control the storage systems that show the purpose of short-term system storage. As a result, the local government may gain system storage without any additional payment. This may ensure a socio-economic benefit of V2G technologies. Moreover, the control and storage systems may prevent the grid electric shock when the current demand gains/loses. In real-time, V2G technology has an infrastructure for grid-response control to predict the demands such as scenario, planning, and data outcome [22].

4.2.1.5 Batteries Permanence

The battery life is so critical and sensitive as it is highly computed upon thermal condition, charge, discharge, etc. As a result, the electric grid may fully control the grid to meet a better battery life. This may influence the socio-economic benefits in advanced metering (AM) and smart grid (SG) to relax the cost of storage systems and electricity. In V2G technology, the energy system is aging a lot in the use of bidirectional to adopt $C/LiFePO_4$ as an alternative to $C/LiNiCoAlO_2$ [23].

4.2.2 Charging Systems

In the electric grid, a bidirectional system supports V2G technology that determines the system efficiency of V2G to cover the core infrastructure. Therefore, a considerable infrastructure may be produced to signify the use of connectors, i.e., Type 1,2,Hybrid-IEC 62196-2, Type 1,2,Combo-Hybrid SAE J1772, and CHAdeMO. The connector namely CHAdeMO is used in DC systems. The energy bases such as renewable energy and power plants convert into DC energies that connect the grid storage networks. When the number of DC lines increases, the THD

TABLE 4.4

Numerical Data Location To V2G Technology

Information Population	US	China	Japan	UK	The Netherlands
EV batteries (k)	404	312	126	49.67	87.53
Electronic charger (k)	28.15	46.65	16.12	8.716	17.78
Fast charger (k)	3.524	12.1	5.99	1.158	0.465
Population (m)	324.6	1373.5	126.8	65.11	17.10
Area Km^2	9,883,520	9,596,961	3,77,972	2,42,495	41,543

ratio may automatically increase the grid computing. In the grid infrastructure, the DC lines may be influential in V2G technology that improves the charging performance in public. The conversion of power electronics may change the energy loss when there is an energy change from DC to AC. The grid infrastructure and THD issue are so critical to reduce the load balance.

Table 4.4 shows the numerical data location for V2G technology. A crucial role in the load balance plays a futuristic challenge to address grid modeling and control. The computation of data accuracy may contribute to predict the future grid. Energy consumption is nowadays efficient to use different energy types that create energy nodes in the urban environment and reserves the DC lines to redistribute the energies, i.e., to form a grid infrastructure [24, 25].

4.2.2.1 Building Charging Station for Renewable Energies

The charging station converts the energy transfer from EV or grid to the storage systems. It is one of the sensible factors to the V2G technologies such as home, office, public place, and parking, i.e., virtual power plant (VPP). It can support high-level energies such as solar energy, CHP, and wind power plant (WPP) to energies at each scale. It shows that the EV batteries may complete the energy circles using V2G technology to perform temporary storage and transmission lines. The office and home are nowadays integrated with V2G and VPP technologies to provide grid energy sectors as shown in Table 4.5 [26].

The major countries supply the charging station as a source of local energy production. It has energy storage and charging system in V2G as a part of the grid source to improve energy efficiencies.

TABLE 4.5

Accessible Distribution in VPP

Energy Location	Home	Office	Parking	Urban
Solar-PV	*	*	*	*
Micro-turbine	*	*		*
Regular-turbine		*		
EV-batteries		*		
CHP	*	*	*	*

4.3   SYSTEM INFRASTRUCTURE

In this section, the strategies, energy pricing, system control and communication, and grid modeling are discussed to signify the energy efficiencies in accord with the target energy and planning scenario.

4.3.1   Strategies

V2G technology works under the target energy and plan. In addition, the target energy has zero-emission, i.e., GHG and efficiency factors such as reliability and stability. Some energy parameters may cause the predetermined changes in the target plan. The energy planning involves energy source, infrastructure, and budget to develop the V2G technologies. It is necessary that the energy source may determine some renewable energy to meet market demands. Therefore, the smart grid, V2G technologies, and renewable energies are driven side by side [27].

4.3.2   Energy Pricing

The driven factors involve the energy sectors to determine the market supply or demand. The grid priorities are given to estimate the energy costs and to balance the overall distribution of the electric grid. In every location, the density of the population cannot be determined, but the specification factor can be inferred by the public or private sector. Therefore, population density, supply/time demand, and location can influence energy pricing [28]. Generally speaking, the location and the population density are supplied to predict the demands of electric grids. In such a condition, the energy loss may increase as the transmission lines are densely occupied. On the other hand, the location may invoke more energy costs as the energy generation is low [29]. Therefore, energy consumption is important to work the electric grid at full capacity, i.e., energy overload.

Figure 4.1 shows the people population in 2016–2050. The current scenario demonstrates that an individual may generate energy for his/her purpose using a renewable source. However, it is completely depending on environmental conditions. Most importantly, he/she can share the energies over a smart grid using V2G technologies. In real time, it may generate the role of a supervisor to balance the energy distribution. Otherwise, the smart grid may cause severe damage to the electric grid. Therefore, the electric grid influences various issues such as applied and cognitive sciences, economy, and psychology to use the EVs. In transportation, people should change into EVs as the fossil energies are depleting to meet the global demands. The industrial forum and the government agencies should organize a systematic group to mitigate the above issue, i.e., to promote the V2G technology.

4.3.3   System Control and Communication

EVs use mobile devices to organize the online duties and observation that may fulfill the supply or demands of the electric grid, i.e., charge/discharge. This may strongly

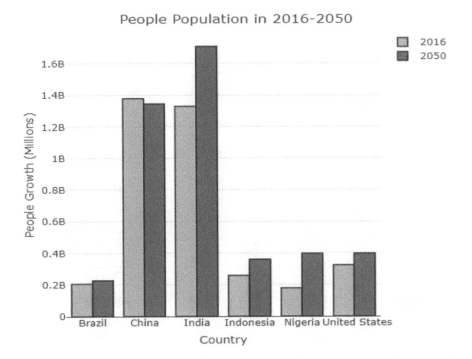

FIGURE 4.1 People population in 2016–2050.

favor the use of modern communication in place of existing charging systems to ensure fine pricing and system management via communication controller, i.e., WAN, HAN, SCADA, GSM, and GPS [30]. Moreover, it has a chain connection to configure the remote data center to address issues related to location and situation (Figure 4.2).

In EVs, data is usually processed to operate the system controller, i.e., in each grid location. However, in a certain situation, the grid may operate abnormally due to power overload or line discharge. Therefore, there may be a method, namely, the ENTRUST algorithm [31] to determine the energy cost, i.e., during the uncertain condition.

4.3.4 SMART GRID MODELING

The smart grid model is based on V2G technology that is driven to achieve better energy efficiency and GHG emission. As a result, there may be a hard situation to investigate the grid regeneration process deeply. As of now, V2G-G2V, renewable energies, and power plants have been utilized as the major input sources. These sources may render the systematic planning and policies to supply renewable energies to define a common subset, i.e., energy efficiencies. It should be tactfully managed to differentiate the energies in particular.

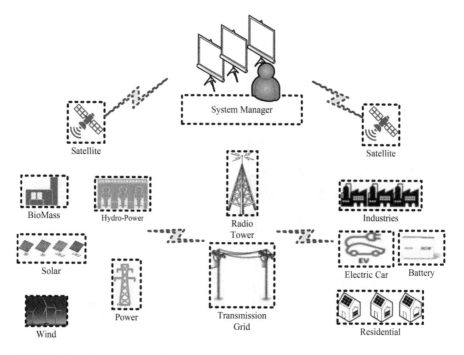

FIGURE 4.2   A system control and communication in electric grid.

4.3.4.1   Frequency Controller

In the grid, power efficiency is one of the influential factors to control the supply or demand when the energies are unbalanced. However, in V2G, the charge and discharge may influence more to determine the grid frequency, i.e., at charging state. EV's batteries with high storage capacities have more energy exchange, i.e., homogenized. Moreover, this cannot be relied on completely due to the imbalance grid frequency. Therefore, the control parameters and grid equipment should be properly quantified to determine whether the EV's capacity is active or not.

4.3.4.2   Power System Integration

A smart unidirectional power system should be designed to connect the grid resources in parallel. The smart design should ensure better storage capacity and limited key issues to meet consumer/industrial demands. V2G technology covers the status of EV's batteries, charging station, and grid capacity. But, the integrated power system may offer less energy loss and high storage efficiency to bridge the gap between the supplies/demands, i.e., in-home or office. Therefore, the infrastructure of the power system should consider the key parameters such as accessibility, reliability, and stability to navigate the energy traffic, i.e., in energy storage systems. This scenario may create systematic group management to recommend a suitable grid infrastructure to fulfill the energy demands.

4.4 CONCLUSION

In the world's electric grid infrastructure, a major revision has been made to address various energy scenarios. The system infrastructure may change upon the enrolment of a new scenario. Moreover, the revision ratio is completely depending on the infrastructure condition and future grid targets. As of now, the energy storage system has become a sensitive part of the developments of micro- and smart grids model. However, the models cannot comply with or satisfy the grid requirements. Therefore, the development of EV's batteries is intended to comply with cyclic systems to ensure better efficiency and safety. Of late, lithium batteries and electronic systems have changed the procedural process of V2G technology, i.e., fast charging mechanism. This evolution may work properly to benefit the electric grid out of energy loss and storage failure. The major key factors are as follows: (1) the cycle systems, i.e., charge cycle or life cycle, energy density, and safety issue, are systematically addressed to improve energy efficiencies; and (2) the grid infrastructure may integrate with a suitable power system to analyze energy traffic.

LIST OF ABBREVIATIONS

CHP:	Combined heat and power
EV:	Electric vehicle
Fe:	Iron
GHG:	Greenhouse gas
GPS:	Global Positioning System
GSM:	Global System for Mobile Communications
G2V:	Grid-to-vehicle
HAN:	Home area network
Li:	Lithium
Mn_2O_4:	Manganese oxide
NaS:	Sodium–sulfur
NiCd:	Nickel–cadmium
PHEV:	Plug-in hybrid electric vehicle
PO_4:	Phosphate oxide
SAE:	Society of Automotive Engineers
SCADA:	Supervisory control and data acquisition
THD:	Total harmonic distortion
VPP:	Virtual power plant
V2G:	Vehicle-to-grid
WAN:	Wide area network
ZnBr:	Zinc–bromine

REFERENCES

1. Loisel, R., Pasaoglu, G., Thiel, C. Large scaled eployment of electric vehicles in Germany by 2030: Ananalysis of grid-to-vehicle and vehicle-to-grid concepts. *Energy Policy* 2014, 65, 432–443.

2. Al-Turjman, F., Abdulsalam, A. Smart-grid and solar energy harvesting in the IoT era: An overview. *Wiley's Concurrency and Computation: Practice and Experience* 2018. DOI: 10.1002/cpe.4896

3. Al-Turjman, F., Abujubbeh, M. IoT-enabled smart grid via SM: An overview. *Elsevier Future Generation Computer Systems* 2019, 96, 1, 579–590.

4. Al-Turjman, F. Modelling green femtocells in smart-grids. *Springer Mobile Networks and Applications* 2018, 23, 4, 940–955.

5. Al-Turjman, F. Mobile couriers' selection for the smart-grid in smart cities' pervasive sensing. *Elsevier Future Generation Computer Systems* 2018, 82, 1, 327–341.

6. Deebak, D., Al-Turjman, F., Mostarda, L. Energy aware resource allocation in multi-hop multimedia routing via the smart edge device. *IEEE Access* 2019, 7, 1, 151203–151214.

7. Mehmood, I., Ullah, A., Muhammad, K., Deng, D., Meng, W., Al-Turjman, F., Sajjad, M., Albuquerque, V. Efficient image recognition and retrieval on IoT-assisted energy-constrained platforms from big data repositories. *IEEE Internet of Things* 2019, 6, 6, 9246–9255.

8. United Nations, Adoption of the Paris Agreement, Conference of the Parties, Twenty-first session, Paris, 30 November to 11 December 2015. Draft decision -/CP.21

9. International Energy Agency, Global EV Outlook. Available online: https://www.iea.org/publications/freepublications/publication/Global_EV_Outlook_2016.pdf (accessed on 28 May 2016)

10. Al-Turjman, F., Altrjman, C., Din, S., Paul, A. Energy monitoring in IoT-based Ad Hoc networks: An overview. *Elsevier Computers & Electrical Engineering Journal* 2019, 76, 133–142.

11. EnergyEfficiency&RenewableEnergy. The Transforming Mobility Ecosystem: Enabling an Energy-Efficient Future. Available online: https://www.energy.gov/sites/prod/-files/2017/01/f34/The%20Transforming%20Mobility%20Ecosystem-Enabling%20a-n%20Energy%20Efficient%20Future_0117_1.pdf (accessed on 29 January 2017).

12. Combined Heat and Power Basics. Available online: https://energy.gov/eere/amo/combined-heat-andpower-basics (accessed on 29 November 2017).

13. Shirazi,Y., Carr, E., Knapp, L. A cost benefit analysis of alternatively fuelled buses with special considerations for V2G technology. *Energy Policy* 2015, 87, 591–603.

14. Lewis, M.F., Amr, S., Francisco, B., Alessandra, S., Deger, S. *Electric Vehicles: Technology Brief; International Renewable Energy Agency*. Abu Dhabi, United Arab Emirates, 2017.

15. Maria, L.T., Michael, L.A. A review of the development of smart grid technologies. *Renewable and Sustainable Energy Reviews* 2016, 59, 710–725.

16. Mathiesen, B., Lund, H., Connolly, D., Wenzel, H., Østergaard, P., Möller, B., Nielsen, S., Ridjan, I., Karnøe, P., Sperling, K., Hvelplun, F. Smart Energy Systems for coherent 100% renewable energy and transport solutions. *Applied Energy* 2015, 145, 139–154.

17. Joy, C.M., Saurabh, S., Arobinda, G. Mobility aware scheduling for imbalance reduction through charging coordination of electric vehicles in smart grid. *Pervasive and Mobile Computing* 2015, 21, 104–118.

18. Zhao, H.R., Wu, Q.W., Hu, S.J., Xu, H.H., Claus, N.R. Review of energy storage system for wind power integration support. *Applied Energy* 2015, 137, 545–553.

19. Opitz, A., Badami, P., Shen, L., Vignarooban, K., Kannan, A.M. Can Li-ion batteries be the panacea for automotive application. *Renewable and Sustainable Energy Reviews* 2017, 68, 685–692.

20. Hsin, W., Edgar, L.-C., Evan, T.R., Clinton, S.W. Mechanical abuse simulation and thermal runaway risks of large format Li-ion batteries. *Journal of Power Sources* 2017, 342, 913–920.

21. Jyri, S., Topi, R., Juuso, L., Peter, D.L. Flexibility of electric vehicles and space heating in net zero energy houses: An optimal control model with thermal dynamics and battery degradation. *Applied Energy* 2017, 190, 800–812

22. Siwar, K., Mouna, R., Lotfi, K.A flexible control strategy of plug-in electric vehicles operating in seven modes for smoothing load power curves in smart grid. *Energy* 2017, 118, 197–208.

23. Harighi, T., Bayindir, R., Hossain, E. Overview of Quality of Service Evaluation of a Charging Station for Electric Vehicle. In *Proceeding of the 2017 IEEE 6th International Conference on Renewable Energy Research and Applications (ICRERA)*, San Diego, CA, USA, 5–8 November 2017, 1180–1185.

24. Kafeel, A.K., Muhammad, A., Saad, M. Inductively coupled power transfer (ICPT) for electric vehicle charging—A review. *Renewable and Sustainable Energy Reviews* 2015, 47, 462–475.

25. Pedro, N., Figueiredo, R., Brito, M.C. The use of parking lots to solar-charge electric vehicles. *Renewable and Sustainable Energy Reviews* 2016, 66, 679–693.

26. Mehdi, A., Mehdi, N., Omer, T. Getting to net zero energy building: Investigating the role of vehicle to home technology. *Energy and Buildings* 2016, 130, 465–476.

27. Farid, A., Esther, P.L., Nathan, W., Bart, D.S., Zofia, L. Fuel cell cars in a microgrid for synergies between hydrogen and electricity networks. *Applied Energy* 2017, 192, 296–304.

28. Liu, L.C., Zhu, T., Pan, Y., Wang, H. Multiple energy complementation based on distributed energy systems—Case study of chongming county, China. *Applied Energy* 2017, 192, 329–336.

29. Morais, H., Sousa, T., Soares, J., Faria, P., Vale, Z. Distributed energy resource management using plug-in hybrid electric vehicle as a fuel-shifting demand. *Energy Conversion and Management* 2015, 97, 78–93.

30. Michael, E., Ramesh, R. Communication technologies for smart grid applications: A survey. *Journal of Network and Computer Applications* 2016, 74, 133–148.

31. Sudip, M., Samaresh, B., Tamoghna, O., Hussein, T.M., Alagan, A. ENTRUST: Energy trading under uncertainty in smart grid systems. *Computer Networks* 2016, 110, 232–242.

5 Data Traffic in IoT-Based Grid

Fadi Al-Turjman
Near East University

CONTENTS

5.1 INTRODUCTION

Recently, heterogeneous wireless technologies have been developed rapidly to support different radio access technologies (RATs) such as GSM, LTE-Advanced, and WiFi in order to connect mobile users to the Internet and form what we call heterogeneous networks (HetNets). In fact, mobile HetNets experience explosive growth in usage and energy consumption due to explosion of smart devices in massive volumes and energy-hungry mobile applications amongst which the smart grid systems start to be the foremost ones [1,2]. Smart grid has been evolved recently in managing our gigantic electricity and energy demands in a sustainable, reliable, and economic manner, while utilizing already existing HetNets' infrastructures. Smart grids are energy networks that can automatically monitor energy flows and adjust to changes in energy supply and demands accordingly. It can be used with smart meters to enhance the quality of experience by providing information on real-time energy consumption. Smart homes can also communicate with the grid using these smart meters and enable consumers to manage their electricity usage. This can be achieved via wireless HetNets connecting the different appliances at home. These home appliances and other electrical devices are heavily used in our daily life and are considered smart devices, and they have also energy-saving issues. The trend in *e-mobility* [3] (i.e., the examined use case in this study), for example, is focused mainly on minimizing the waste in electrical energy provided by the municipal utility. Smart grids in such scenarios can improve operating efficiencies, lower costs, shorten outages, and reduce peak demands and electricity consumption. This is due to automated meter readings, and service connections and disconnections which can significantly reduce the need to dispatch trucks and personnel while accomplishing these tasks [4], where reducing the truck rolls by 15,000 miles can save approximately 6.3 metric tons of CO_2 emissions. Also the integrated wireless communication networks in smart grids can be used for outage management, where utilities can ping smart meters to determine which customers don't have power and pinpoint outage locations for quicker and more efficient service restoration. In addition, it can accommodate future smart grid upgrades, automate other city services (such as water and gas metering), and deliver valuable returns for cities and taxpayers. Accordingly, the utilized HetNets by the municipal of the city is used not only to connect, communicate, and control most of the city's smart grid equipment/devices but also to avoid the global energy waste and reduce the carbon footprint. In order to realize these benefits, the utilized HetNets have to overcome several challenges in practice. These include integrating heterogeneous technologies and communication systems optimally for data backhauling, catering with the mobility factor, tolerating system failures to reduce electricity consumption and lower bills, and implementing new communication systems in a dynamic environment of rapidly evolving users and devices. Towards this end, there is a need for an efficient HetNet between the different heterogeneous appliances/devices at home and outside the home such as the home area network (HAN) and the neighborhood area network (NAN). HAN consists of three components: it measures, collects, and analyzes energy usage from smart devices [5]. NAN connects multiple HAN to local/regional access points (i.e., access points within the range of instantaneous transmitters), where transmission lines carry the data to the municipal [5]. Such HetNets in the smart grid need a communication approach that connects the vast counts of consumers and suppliers in an energy-efficient way.

One promising solution for HetNets and cellular providers in this regard is the deployment of femtocells [2]. A femtocell is a cell which provides cellular coverage and is served using a femtocell base station (FBS), which is short-range and low-power cellular base-station typically deployed in indoor environments and/or outdoor rural and densely populated areas for enhanced reception of voice and data traffics [6,7]. It broadcasts using GSM signals and allows a cellular phone to connect to the Internet for better indoor coverage. It uses between 8 and 120 mW, which is much lower than a WiFi access point [4]. And thus, it can be used as a green communication method in HAN and HetNets in general, where consumers can save energy and money using their smart home systems. They can track their energy usage, and this can give consumers the ability to manage their electricity bills. Moreover, with the help of the femtocell, appliances can be automated and remotely controlled to reduce unnecessary electricity demands on the grid. This can be generalized for all smart homes in a smart city scenario, for example, where mobile outdoor femtocells can be utilized to achieve ultra-large-scale (ULS) coverage. However, this increases the HetNets overload and makes our expectations unrealistic in proximity of the green planet vision [2].

In fact, RATs are responsible for more than 2.5% of the total carbon emission rates, which is more than the carbon emission rate of the global aviation industry [4,8]. This rate is expected to be doubled over the next decade [4], which means more energy waste since carbon emission is a straightforward result of energy usage in the world, especially when we know that the main component of a HetNet, which is the macro-base station, consumes between 2.5 and 4 kW [4]. The number of these stations in every country is about tens of thousands and still growing more and more. This means lots of energy consumption demands and carbon emissions are coming. It is estimated that only between 5% and 10% of this energy will be used to create useful signals. That is because macrocells provide wide area coverage and most of the area is empty space where transmitted signals are not utilized. Moreover, the radiated wireless signals need high power without efficiency, which means power is used for just in case scenarios where a user might exist. And thus, femtocell is a powerful candidate to prevent these energy losses. It provides local coverage while macrocell is providing huge area coverage. In other words, femtocell delivers power only where there is a need. So it consumes less energy to provide higher bandwidth since it is closer to user. The other advantage is that user equipment (UE) consumes less energy when it is connected to femtocell since it is closer, where it takes 40 times more energy to deliver signal to indoor from a macrocell compared to a femtocell. Thus, femtocell is more energy-efficient in addition to providing more coverage and capacity, and better quality of service (QoS). Nevertheless, significant energy amounts can be wasted in data (re)transmission unless a reasonable load balance is applied between the deployed FBSs which are typically planned to serve huge counts of static/mobile users of the smart grid. This would not be achieved without a realistic case study analysis and an accurate analytical model that can predict the system performance in such setup.

There have been a few attempts in the literature to analytically model the energy consumption in a femtocell. On-off model is the most basic one which can be used for theoretical analysis where FBS is assumed to consume unit power in active mode and zero power when it is off. However, it does not reflect actual power consumption. In Ref. [9], the authors proposed a linear power model that considers traffic load. This model is used for analysis and more accurate than on-off model since traffic load is considered. In Ref. [10], the authors proposed a simple analytic model to predict the FBS power consumption based on the offered load and datagram size. However, they neglected that radio energy must be consumed by the base station (BS) when making downlink transmission. In Ref. [11], energy consumption is modeled based on three main interactive components in the FBS: the microprocessor, the FPGA, and the radio-frequency transmitter for indoor applications. Nevertheless, it is mainly designed for static indoor applications and cannot predict outdoor energy consumptions for mobile FBS.

In this study, unlike the aforementioned studies, an analytical modeling approach is proposed, where varying work load, communication range, and multiple FBS/ user mobility-related issues as well as channel failures in the femtocell infrastructure are considered [12]. Our modeling approach has been utilized in a quite useful offloading approach for discovering the operational space of various femtocell configurations/decisions. This approach is maintained mainly in two algorithms. It leads to more accurate QoS measurements in real-time applications (e.g., voice over IP (VoIP) and video streaming), while considering the varying speed effect of the mobile FBS/ UE on the HetNet performance of a smart grid and other QoS parameters. Unlike other approaches in the literature, our approach caters for failures and data packets which can leave the system due to mobility and energy-saving modes. And the effect on the performance characteristics of the grid system such as mean queue length (MQL), throughput, and delay has been investigated accordingly. In the following, the main contributions of this research are summarized:

- A queue-based analytical model is proposed while considering varying characteristics in the grid infrastructure.
- Femtocell predictions in energy consumptions with high accuracy and efficacy have been achieved and discussed.
- A novel offloading approach for discovering the operational space of various femtocell configurations/decisions is proposed.
- A hybrid wireless cellular HetNet consisting of a macrocell and several femtocells has been considered as a case study towards more energy-efficient HetNets in the smart grid era.
- Detailed analysis of the case study is given based on the queuing theory concept, and results have been validated through extensive simulations.

The remainder of this chapter is organized as follows. Section 5.2 overviews the related attempts for HetNet modeling in the literature. Section 5.3 highlights the assumed and used system models. In Section 5.4, we propose our detailed grid model towards realizing green femtocell applications. Next, a real case study, namely, the *e-mobility*, is examined and used to validate our proposed model via extensive simulation results in Section 5.5. And finally, we conclude this work in Section 5.6.

5.2 RELATED WORK

There are several attempts in the literature towards modeling HetNets telecommunication systems for better performance assessments and energy consumption predictions. Modeling these systems using analytical models like queuing theory is a well-known approach in the literature [11,13–17]. Different performance metrics, depending on system and required analysis' studies, such as the average number of requests in the system, average resource utilization, average power consumption, average waiting time, and throughput, have been investigated [15,16]. Such modeling systems and performance metrics can be classified into static and dynamic models. By static and dynamic models, we refer to systems with/without mobile femtocells. In static models such as the ones presented in Refs. [18–20], performance characteristics of cellular networks have been investigated without considering mobility, and thus, we call it static systems. Unlike static models, in the dynamic models, mobility is considered as one of the utmost important issues in the performance evaluation process [21–27].

5.2.1 STATIC MODELING

Several studies have been performed on analyzing and evaluating the performance of typical small cell deployments in static scenarios (i.e., without mobility considerations). For example, in Ref. [13], a set of algorithms have been proposed to reduce the energy consumption of a dense network and provide better QoS to the end-users. In Ref. [13], the authors studied the energy consumption of a campus WiFi. A simple approximation queuing model is used to save energy in WiFi by considering cut-offs for the small cells according to user demands due to sleep modes, channel failures, mobility issues, etc. Presented results show that by using sleep modes for the small cell, a considerable amount of energy can be saved when the number of users connected to the network is small. In Ref. [17], the authors evaluate the performance advantage of using fixed small cells as relays by communicating with the macro-BS to improve and extend the HetNet coverage. In Ref. [11], the authors proposed a simple analytical model to predict a static FBS power consumption based on a specific offered load and packet size. They assumed one femtocell that supports up to four simultaneous end-user devices in a campus network to predict energy consumption of voice and FTP messages. However, in their prediction, they neglected that radio energy consumption while performing downlink transmissions. In Ref. [19], the authors analyzed the behavior of adaptive modulation and coding (AMC) systems with sleep mode using queuing theory. They are interested in evaluating energy consumption rates per packet, average delays, and packet loss. An admission control problem for a multi-service LTE radio network is addressed in Ref. [20]. A model for two resource-demanding video services, video conferencing and video on demand, is proposed. Teletraffic and queuing theories are applied to obtain a recursive algorithm in order to calculate performance measures such as blocking probability and the mean bit rate. However, a limited number of researchers have studied and modeled mobile small cell deployments as an option in providing green systems.

5.2.2 DYNAMIC MODELING

In Ref. [21], the authors propose the idea of deploying small cells in vehicles to improve the uplink throughput for mobile users. Results show that mobile small cells can enhance QoS and maintain an acceptable level of signal-to-interference noise ratio (SINR). In Ref. [22], the authors proposed seamless multimedia service for mobile users in high-speed trains through deploying small cells on board. The onboard small cells communicate with macrocells to facilitate the seamless handover. In Ref. [23], an integrated cellular/WiFi system is modeled for high mobility using a two-stage open queuing system with guard channel and buffering to obtain acceptable levels of QoS in heterogeneous environments. An exact analytical solution of the system is given using the spectral expansion solution approach that can be useful for vertical handover decision management. Similarly, in Ref. [24], the authors model an integrated cellular/WiFi HetNet in order to study specific performance characteristics such as MQL, blocking probability, and throughput. The system is modeled as a two-stage open queuing network, and the exact solution is presented using the spectral expansion solution approach. Simulation is also used to validate the accuracy of the proposed system. In Ref. [24], the authors presented a mathematical model for analytical study on complete and partial channel allocation schemes. By using Markov models, results can be presented for performance measures such as MQL and blocking probability.

However, amongst the most significant issues in performance evaluation of such HetNets is mobility [26]. In Ref. [26], wireless cellular networks are modeled using a Markov reward model. An S-channel per cell in homogeneous cellular system and mobility-related issues are considered. Performance characteristics of the system such as MQL and blocking probability are presented using an analytical model. In Ref. [27], the authors considered a network in which cells of different sizes have been deployed depending on mobile user density, traffic, and coverage such that power consumption can be minimized without compromising QoS. They developed analytical models of power consumption in five different ways and obtained the reduction in power consumption compared to macrocell networks. In the first one, they used femtocell-based network instead of the macrocell-based one in an area that is fully covered by femtocells only and accordingly obtained between 82.72% and 88.37% power consumption reduction. In the second way, they divided the area into three parts as urban, suburban, and rural areas. They considered mobile user density, mobile user traffic, and required coverage, and covered urban areas by femtocells, suburban areas with macrocells, and rural areas with mobile femtocells. As a result of this setup, they achieved between 78.53% and 80.19% reduction in power consumption. In the third way, they allocated femtocells to densely populated urban area, picocells to sparsely populated urban areas, microcells to suburban areas, and mobile femtocells to rural areas, and the reduction in power consumption rate was predicted to be between 9.19% and 9.79%. In the fourth way, they allocated microcells, picocells, and femtocells to the border region and macrocell to remaining region. The reduction in power consumption is between 5.52% and 5.98% for this setup. The last one, femtocells are allocated at the boundaries of macrocells, where the radio signal is not enough for making a call. As a result of the last setup, a reduction in power consumption between 1.94% and 2.66% and a macrocell coverage shrink is achieved. Moreover, two different handoff schemes with/without preemptive priority procedures for integrated wireless mobile networks are proposed and analyzed in Ref. [28]. Service requests are categorized into four different types: as voice requests, data requests, voice handoff requests, and data handoff requests. A 2D Markov chain is used to model the system and analyze the HetNet performance in terms of average delay, blocking probability, and forced termination probability. Existing research efforts, however, do not target assessing the performance gains of mobile small cells. A quantitative performance analysis of such gain is definitely needed nowadays for better energy utilization and more green applications. And hence, the work in this chapter is proposed. Particularly, we consider the different velocity effect of the modeled mobility which is a typical case in smart grid applications such as the *e-mobility* project in Siemens [3].

5.3 SYSTEM MODELS

In this study, mobile users may move to neighboring cells while they are either in the queue or being served in the system. It is typical in grid systems to experience some outage periods due to many different reasons including load balance and/or sleep modes for better energy consumption. These cut-offs and unavailability of an FBS may degrade the performance of the grid system. It is assumed that a single recover facility is available not only for every FBS, but also for every FBS channel to make the cut-off recovered again. Similar to previous studies in Refs. [7,21], each macrocell can be represented by a circle of radius R so that it is served by a BS placed at the center. The femtocells which are deployed within the coverage area of a macrocell are also represented by circles of radius r and are served by mobile/static FBSs.

5.3.1 QUEUING MODEL

In order to satisfy energy requirements in grid-based HetNets applications, a Markovian discrete-time stochastic process $M/M/N/L$ queuing model is assumed. To cope with the grid heterogeneity nature, we assume a priority-based approach in queuing the incoming requests. The queuing capacity of the system is denoted by W, and L represents the maximum capacity which includes the number of FBSs in the system (N). Similar to related studies in Refs. [28–30], arrivals to the system are assumed to follow a Poisson process with arrival rate σ, and service time exponentially distributed with rate μ. This system is proposed under a realistic assumption of a mobile FBS queue that can hold waiting requests as long as they are within the required communication range. The arrival and departure of the data packets are regulated under a finite queue size. Multiple mobile FBSs may move while they are inside the coverage area of a macrocell, and thus, user requests will handover to other neighboring cells with rate μ_{cd} while they are either in the queue or being served in the system. Moreover, cut-offs may also occur in the system. The cut-off rate of the FBS is assumed to be exponentially distributed and is denoted by ξ [28]. Following the cut-off, the failed FBS is recovered with a recover rate η with exponential distribution as well. Figure 5.1 represents our FBS queuing system under this study. Symbols/notations used in this model are summarized also in Table 5.1. According to Ref. [16], the dwell time of a mobile FBS is the time that the mobile node spends in a given system. The dwell time is assumed to have an exponential distribution with a mean rate μ_{cd}. And thus, the service rate due to mobility can be calculated by

$$\mu_{cd} = \frac{P.E[v]}{\pi A},$$ where $E[v]$ is the average expected velocity of the mobile FBS. P and A are the length of the perimeter and the area of the macrocell, respectively.

5.3.2 COMMUNICATION MODEL

Practically, the signal level at distance d from a transmitter varies depending on the surrounding environment. These variations are captured through what we call

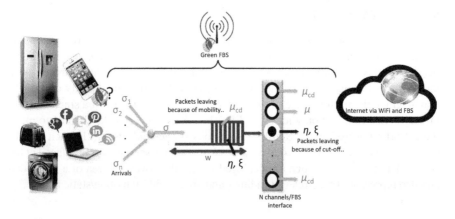

FIGURE 5.1   The queuing system considered with cut-off and mobility effects.

TABLE 5.1
Summary of Symbols

Symbol	Definition
R	Radius of the femtocell
V	Velocity of the mobile users
P	Perimeter of the femtocell
A	Area of the femtocell
N	Total number of channels in the FBS
W	Queue capacity of the cell
L	Maximum number of requests in the cell
σ	Total arrival rate of requests in the cell
μ	Total service rate of completed request departures in the cell
μ_{cd}	Mean service rate of handover requests in the cell
ζ	Cut-off rate of a server
η	Mean recover rate of the cut-off

log-normal shadowing model. According to this model, the signal level at distance d from a transmitter follows a log-normal distribution centered on the average power value at that point [31]. This can be formulated as follows:

$$P_r = K_0 - 10\rho \log(d) - \chi \tag{5.1}$$

where d is the Euclidian distance between the transmitter and receiver, ρ is the path loss exponent calculated based on experimental data, χ is a normally distributed random variable with zero mean and variance σ^2, i.e., $\chi \sim \mathcal{N}(0, \sigma^2)$, and K_0 is a constant calculated based on the mean heights of the transmitter and receiver.

5.3.3 ENERGY CONSUMPTION MODEL

Assuming d is the distance in meters between the transmitter in a smart grid point of interest and the mobile FBS. The achievable transmission rate at the FBS can be approximated using Shannon's capacity equation with signal-to-noise ratio (SNR) clipping at 20 dB for practical modulation orders as follows:

$$R = B \log_2 \left(1 + \frac{P_r}{N_0.B}\right) \qquad (5.2)$$

where R is the data rate in bit per second for a given received power P_r and system bandwidth B. The P_r is computed using Eq. (5.1), and N_0 is the background noise power spectral density (i.e., Additive white Gaussian noise (AWGN)). Consequently, and based on Ref. [32], the total power consumption, P_{tot}, of an FBS can be formulated as $P_{tot} = P_{FPGA} + P_{mp} + P_t + P_a$, which is the total of power consumption parameters of the FPGA, the microprocessor, the transmitter, and the power amplifier, respectively. And thus, we define the energy consumed at a FBS by $E_{FBS} = P_{tot} * R * (E_{TX} + E_{RX})$, where most of the energy consumption at the FBS is due to data communication, indicated by E_{TX} for transmission energy and E_{RX} for energy consumed during the data reception process. And T represents the number of transmitted packets. And hence, expected energy consumption at a FBS with communication N channel/interface can be estimated as follows:

$$E(x) = \sum_{i=1}^{N} P_i.i.\mu.E_p \qquad (5.3)$$

where P_i, i, μ, and E_p are probability of having i channels available (sum of all probabilities in columns of the 2D state diagram shown in Figure 5.2), number of available channels, service rate, and energy consumption for each transmitted packet, respectively. In this work, we refer to one-hop neighbors' communication as the first tier of nodes. Since no other node can reach the macrocell station directly, traffic from every other node will have to be forwarded, in the last hop, by one of the FBSs. If the spatial distribution of nodes is assumed to be uniform, then the traffic load is equally distributed. Each first tier node will forward hardly the same amount of traffic, and all first tier nodes' energy will be depleted at times very close to each other, after the network is first put into operation. Since all of the first tier nodes' energies are depleted at once, the FBS will be overwhelmed and plenty of handovers will be performed, in addition to experiencing peak energy-consumption periods. Increasing the number of nodes in the network accentuates this effect, since there is more traffic to forward and the first tier of nodes has no indicator to manage their energy budget. And hence, delegating the role of device scheduling to the FBS based on the proposed energy consumption model can lead to significant gain in terms of energy savings and quality of service.

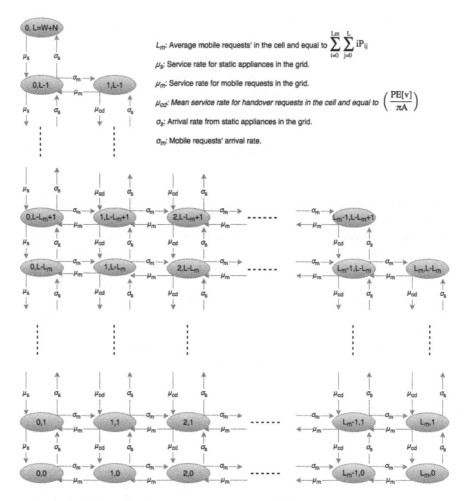

FIGURE 5.2   The 2D Markov chain for the targeted grid system.

5.4   THE GREEN FBS MODEL FOR SMART GRIDS

Future smart grids are expected to be a combination of macrocells and small cells such as femtocells [33]. FBSs are mostly expected to be static/mobile stations on buses, taxis, trains, smart homes, etc. in a grid application such as Siemens' *e-mobility* [3] to provide better coverage and capacity to dramatically increasing mobile users. The overall system can be considered in Q-theory as a 2D Markov process. A pair of integer valued random variables, i and j, can be used to describe the state of the system at time t, where $i(t)$ represents the number of available communication channels/interfaces at time t, and $j(t)$ determines the number of existing requests/packets at time t. A 2D state diagram for the Q-system is considered in this study as depicted in Figure 5.2. We can assume the minimum value of i is 0, and the maximum is N representing the maximum number of channels in the system. For the random variable j, the minimum value is 0 and it can take values from 0 to L which is the total number of requests in the system at time t, including the one(s) in service (i.e., $W + N$). The Markov process is denoted by Z and is used for performance evaluation of the considered grid system in this chapter. We assume Z is irreducible with a state space of $N\times N$. Furthermore, we assume that the number of channels/FBS interfaces, $i(t)$, is represented in the horizontal direction and the total number of requests, $j(t)$, is represented in the vertical direction of a finite lattice strip. And thus, possible transitions of the grid system Z are purely lateral transitions from state (i, j) to state (k, j), one-step upward transitions from state (i, j) to state $(k, j+1)$, and one-step downward transitions from state (i, j) to state $(k, j-1)$. In this study, spectral expansion approach is used where A is the matrix of purely lateral transitions with zeros on the main diagonal, and one-step upward and one-step downward transitions are represented in matrices B and C, respectively.

$$A = \begin{pmatrix} 0 & \eta & 0 & 0 & 0 & 0 & 0 & 0 \\ \xi & 0 & \eta & 0 & 0 & 0 & 0 & 0 \\ 0 & 2\xi & 0 & \eta & 0 & 0 & 0 & 0 \\ 0 & 0 & 3\xi & 0 & \ddots & 0 & 0 & 0 \\ 0 & 0 & 0 & \ddots & 0 & \ddots & 0 & 0 \\ 0 & 0 & 0 & 0 & \ddots & 0 & \eta & 0 \\ 0 & 0 & 0 & 0 & 0 & (n-1)\xi & 0 & \eta \\ 0 & 0 & 0 & 0 & 0 & 0 & n\xi & 0 \end{pmatrix} \tag{5.4}$$

$$B = \begin{pmatrix} \sigma_s,\sigma_m & 0 & 0 & 0 & 0 & 0 & 0 \\ 0 & \sigma_s,\sigma_m & 0 & 0 & 0 & 0 & 0 \\ 0 & 0 & \sigma_s,\sigma_m & 0 & 0 & 0 & 0 \\ 0 & 0 & 0 & \ddots & 0 & 0 & 0 \\ 0 & 0 & 0 & 0 & \ddots & 0 & 0 \\ 0 & 0 & 0 & 0 & 0 & \sigma_s,\sigma_m & 0 \\ 0 & 0 & 0 & 0 & 0 & 0 & \sigma_s,\sigma_m \end{pmatrix}, \tag{5.5}$$

$$C = \begin{pmatrix} \min(0,j)\mu_s, \mu_m + j\mu_{cd} & 0 & 0 & 0 & & 0 \\ 0 & \min(1,j)\mu_s, \mu_m + j\mu_{cd} & 0 & 0 & & 0 \\ 0 & 0 & \ddots & 0 & & 0 \\ 0 & 0 & 0 & \ddots & & 0 \\ 0 & 0 & 0 & 0 & & \min(n,j)\mu_s, \mu_m + j\mu_{cd} \end{pmatrix},$$

$$0 \le j \le L$$

$$(5.6)$$

In this grid system, the transition rate matrices are always dependent on j, because the requests in the queue may leave the system due to the assumed user/FBS mobility.

Lemma 5.1

Elements of matrix A depend only on the cut-off and recover rates of the FBS channels, ξ and η, respectively. The transition rate matrices A, B, and C are square matrices each of size $(N+1)\times(N+1)$ and given in Eqs. (5.4)–(5.6). The matrix C depends on the number of requests in the system for $j = 0$, 1, ..., L. And thus, the threshold M of the spectral expansion approach can be set to L.

If the number of requests in the system is less than the number of available channels, each request is served using a channel. On the other hand, if the number of requests is greater than the number of available channels, N requests with the highest priority are served first and the remaining can only handover to a neighboring cell with the service rate μ_{cd}. The spectral expansion solution approach is used for the steady-state solution. Details of the spectral expansion solution approach can be found in Ref. [34]. Following spectral expansion solution, Lemma 5.2 can be stated as follows.

Lemma 5.2

The steady-state probabilities of the states can be expressed as

$$P_{ij} = \lim_{t \to \infty} \left(P(i(t) = i, \; j(t) = j), \quad 0 \le i \le N \text{ and } 0 \le j \le L \right) \tag{5.7}$$

From *Lemma 5.1* and *Lemma 5.2*, it is possible to obtain *Theorem 5.1* as follows:

Theorem 5.1

Probability of being in a state ij is

$$P_{ij} = \sum_{l=0}^{N} \left(a_l \psi_l(i) \lambda^{j-M+1} + b_l \varphi_l(i) \beta^{L-j} \right), \qquad M-1 \le j \le L \tag{5.8}$$

where, λ and ψ are eigenvalues and left-eigenvectors of $\varphi(\lambda)$, respectively, and ψ is a row-vector defined as $\psi = \psi_0, \psi_1 \ldots \psi_N$, λ is a row-vector defined as $\lambda = \lambda_0, \lambda_1 \ldots \lambda_N$, and $\psi\varphi(\lambda) = 0$; $|\varphi(\lambda)| = 0$. On the other hand, β and φ are eigenvalues and left-eigenvectors of $\underline{Q}(\beta)$, respectively, and φ is a vector defined as $\varphi = \varphi_0, \varphi_1 \ldots \varphi_N$, $\beta = \beta_0, \beta_1 \ldots \beta_N$.

Consequently, the state probabilities in *Theorem 5.1* can be used to calculate important performance measures such as MQL, response time (R), and throughput (γ) as follows:

$$MQL_m = \sum_{i=0}^{L_m} \sum_{j=0}^{L} iP_{ij}, \qquad MQL_s = \sum_{i=0}^{L_s} \sum_{j=0}^{L} iP_{ij}$$

$$\Rightarrow MQL = MQL_m + MQL_s \qquad (5.9)$$

$$\gamma_m = \sum_{i=0}^{L_m} \sum_{j=0}^{L} \mu_m P_{ij}, \qquad \gamma_s = \sum_{i=0}^{L_s} \sum_{j=0}^{L} \mu_s P_{ij}$$

$$\Rightarrow \gamma = \gamma_m + \gamma_s \qquad (5.10)$$

$$R_m = \frac{MQL_m}{\gamma_m}, \qquad R_s = \frac{MQL_s}{\gamma_s}$$

$$\Rightarrow R = R_m + R_s \qquad (5.11)$$

Base on the aforementioned formulas, the FBS simply can decide on whether to accept a new user in the queue or not, where a UE should communicate with the FBS that has the best SNR. In fact, the UE_i will report the mobile FBS's SNR to the serving macrocell and send a Req_i to offload. Algorithm 5.1 represents this stage at the macro-BS. When the SNR trigger condition is satisfied, the serving macro-BS checks the status of the UE indicated in lines 2–5. If UE's status is active, the macro-BS will classify the UE_i traffic as indicated in line 6. As a result, the user category (C_i) will be associated with UE_i. The macro-BS waits a predefined residence time (res_time) and then sends the Req_i with its associated C_i to the mobile FBS as in lines 7–8. However, if UE_i status is other than active, the macro-BS ignores its request and keeps it associated with macro-BS. After receiving the decision from the mobile FBS, the macro-BS will switch (offload) the UE_i to the mobile FBS if it receives an *Accept* message from the FBS (lines 8–9), where the *Decide()* function can be as detailed in Algorithm 5.2.

Based on Algorithm 5.2, once the mobile FBS receives the Req_i from an UE_i, it checks whether to accept or not (lines 2–12). If the FBS chooses to accommodate the UE_i, it informs the serving macro-BS to transfer the data session of UE_i and update the list of offloaded users (U_f), and their total counts $T_{total,}$ (lines 3–4). However, if the FBS reaches its maximum threshold L, where L here is equal to U_{fmax}, and there exists an UE_i connected with the same FBS but with lower priority than the new UE_j. In this case, the mobile FBS will transfer/return the user with the minimum priority (UE_{min}) to the macro-BS and accept the new UE_j (lines 6–9). Finally, if the

Algorithm 5.1: UE$_i$ Categorization at the Macro-BS.

Input: *Req$_i$*: a request by UE$_i$ to switch to mob-FBS
1. Receive a Req$_i$ from a UE$_i$
2. **If** UE$_i$ is active **then**
3.   **If** the UE$_i$ has a voice call **then**
4.   Ignore // i.e. keep connected to the macro-BS
5.   **Else**
6. C_i = Classify UE$_i$ based on application types
7.   wait for *res_time*
8.   **If (Decide** (Req$_i$) $==$ Accept) **then**
9.   Transfer UE$_i$ to the FBS
10.   **Else**
11.   Ignore
12.   **Endif**
13. **Endif**
14. **Else**
15. Ignore
16. **Endif**
17. **End**

Algorithm 5.2: Decision at the Mobile FBS.

Decide (Req$_i$)
Input: Req$_{i:}$ is a data user request from a UE$_i$ associated with its C_i
Output: <u>Accept/Reject</u>: message sent to macro-BS to transfer/keep the UE$_i$
Initialize: U_f, U_{fmax}, $T_{total,}$ UE$_{min}$
1. Receive a Req$_i$ from the macro-BS of UE$_i$
2. **If** $U_f < U_{fmax}$ AND $T_{total} \leq L$ **then**
3.   Accept UE$_i$
4. $U_f = U_f + UE_i$
5. $T_{total} = T_{total} + 1$
6. **Elseif** $U_f = U_{fmax}$ AND $\exists$ C$_{i-1}$ $\in U_f$ **then**
7.   Switch UE$_{min}$ to macro-BS & Accept UE$_i$
8. $U_f = U_f + UE_i$
9. $T_{total} = T_{total} + 1$
10. **Else**
11.   Reject UE$_i$
12. **Endif**
13. **Endif**
14. **If** FBS is cut-off **then**
15.   Transfer $\{U_f\}$ back to the macro-BS
16. **Endif**
17. **End**

WiFi signal strength degrades below a certain threshold, the mobile FBS will ask the macro-BS to take its list of UE (lines 14 and 15).

5.5 A TYPICAL CASE STUDY IN SMART GRIDS: *E-MOBILITY*

e-Mobility is a real case scenario proposed and instantiated by Siemens [3] for applied mobile FBS in smart grids not only as a mean of transportation, but also to feedback electricity into the grid at the peak hours. It represents the concept of using electric powertrain technologies, in-vehicle information, and communication technologies and connected infrastructures to enable the electric propulsion of vehicles and fleets. Powertrain technologies include full electric vehicles and plug-in hybrids, as well as hydrogen fuel cell vehicles that convert hydrogen into electricity. e-Mobility efforts are motivated by the need to address corporate fuel efficiency and emission require-ments, as well as market demands for lower operational costs. Electrical vehicles/trains in this case study are relying on mobile FBS in exchanging their energy status during the day in addition to other smart home appliances, and thus, a heavy data traffic is expected to be generated. In such a comprehensive mobile HetNet model, an efficient energy consumption policy is required to motivate the usage of femtocells in serving thousands of incoming requests per hour in a green framework. Moreover, such kind of a mobile HetNet model introduces plenty of challenges regarding the system's capacity and targeted QoS. Hence, we visualize a green HetNet-driven fem-tocell case study for *e-mobility* in smart grid that tackles the aforementioned concerns.

We consider a set of femtocells which are deployed inside the coverage area of a macrocell to provide sufficient users' capacity while maintaining adequate QoS in terms of throughput, MQL, response time, and energy consumption. Mobile users might be static/mobile and may use their smartphones and energy-hungry mobile applications such as mobile video streaming while they are commuting over the city road. Each FBS can be described via a set of hardware parameters as described in Section 5.3. Typically, users are assumed to be uniformly distributed in the cover-age area of their serving cell. The FBS parameters used in the proposed queuing system for this case study are summarized in Table 5.2 while assuming typical LTE-values that have been used in practice [35–37], and the assumed *e-mobility* scenario is shown in Figure 5.3. In order to assess the proposed Q-model under this scenario, we consider the summarized performance metrics in Table 5.3.

5.5.1 SIMULATION SETUPS

Using MATLAB® R2016a and Simulink® 8.7, we simulate randomly generated heterogeneous networks to represent the targeted smart grid environment in a smart city. A discrete event simulator is built on top of these MATLAB platforms which con-siders practical aspects in the network physical layer for more realistic performance evaluations. Our *Simulink* simulator supports wireless channel temporal variations, node mobility, and cut-offs. Based on experimental measurements taken in a site of dense heterogeneous nodes [38], we adopt the described signal propagation model in Section 5.3, where we set the communication model variables as shown in Table 5.2, and χ to be a random variable that follows a log-normal distribution function with

TABLE 5.2

Specifications of an FBS Parameters [32,35]

Parameter	Value
P (mW)	20
BW (MHz)	5
N_0 (W/Hz)	$4*10^{-21}$
p_{mp} (W)	3.2
p_{FPGA} (W)	4.7
p_{trans} (W)	1.7
p_{amp} (W)	2.4
FBS radius (m)	30
FBS velocity (km/h)	Low, medium, high
FBS channels	8
Expected cut-off rate per hour (ξ)	0.001
Expected arrival rate per hour (σ)	2000
Expected recover rate per hour (η)	0.5
ρ	4.8
δ^2	10
K_0	42.152
R	30 m

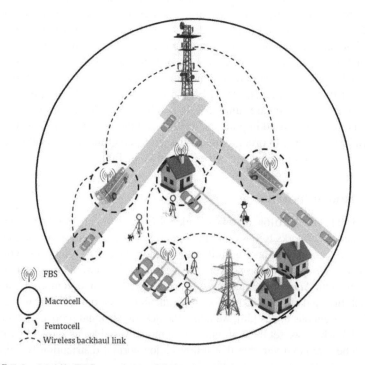

((ᵗᵖ)) FBS

◯ Macrocell

⟨ ⟩ Femtocell

- - - Wireless backhaul link

FIGURE 5.3 Mobile FBSs serving mobile/static users in a smart grid of FBSs.

TABLE 5.3
Performance Metrics and Parameters

Performance Metrics and Parameters	Definitions
Throughput (γ)	The average percentage of transmitted data packets that succeed in reaching the destination. This metric has been chosen to reflect the effectiveness of the mobile FBS in a HetNet setup, and it is measured in "*packets/h.*"
MQL	The average number of the requests pending in the system, either waiting in the queue or being served. This metric represents the QoS from the FBS perspective and is measured in "*packets.*"
Response time (R)	The time spent by a mobile/static user from arrival until departure and plays a significant role in performance evaluation since it incorporates all the delays involved per user request.
Energy consumption (E)	The amount of energy consumed by a single FBS based on the arrived user requests and the FBS dynamic status change.
Service rate (μ)	The number of served users per hour at a FBS. It is used to analyze the effect of traffic loads on the performance of the FBS system.
FBS velocity	The velocity of a mobile FBS measured in km/h.

mean 0 and variance of δ^2. In this simulator, an event-based scheduling approach is taken into account which depends on the events and their effects on the system state. The assumed event-based scheduling approach is typical for data gathering in wireless networks. It allows multiple, parallel indirect transmissions across multiple, adjacent clusters, with collision avoidance techniques. Since we are assuming priority-based queue, scheduled packets can be delayed in favor of the node with the highest rank, and packets having the same sender are ordered according to their arrival time. As for the stopping criterion, we assume a commonly used one, called relative precision [9]. It is used in our simulations to be stopped at the first checkpoint when the condition $\beta < \beta_{max}$, where β_{max} is the maximum acceptable value of the relative precision of confidence intervals at $(1-\alpha)$ significance level. Accordingly, our achieved simulation results are within the confidence interval of 5% with a confidence level of 95%, where both default values for β and α are set to 0.05. The simulation results obtained from our MATLAB code are presented comparatively with the analytical results from our queue model and validated to reflect the performance of the actual femtocell system.

5.5.2 Results and Discussions

Obtained results for the proposed case study are presented in this section in two subsections. First, impact of the mobility-speed factor on an FBS performance in terms of throughput and queue length is studied. Second, the impact of traffic load and service rate on the FBS energy consumption is considered and analyzed.

5.5.2.1 The Impact of Velocity on FBS Performance

Based on the arrival rate (σ) that varies in this study from 2500 to 6000 users/h, performance metrics are compared for three different FBSs' mobility speed categories: (1) low-speed FBSs such as stationary ones at smart homes and in pedestrian handheld smart devices with the velocity from 0 to 15 km/h, (2) medium-speed FBSs like those on top of buses with the velocity from 15 to 40 km/h, and (3) high-speed FBSs with the velocity above 40 km/h, such as those on top of electric cars/trains.

Figure 5.4 shows the effect of velocity of the mobile FBS on the MQL for various arrival rates. It is clear from the figure that when the system is congested like in densely populated areas such as airports and city centers, the MQL will also grow. This is because more users request service from the FBS at the same time. As the mobile FBS moves faster, the MQL decreases. This is due to the fact that the service rate, μ_{cd}, is directly proportional to the expected velocity of mobile FBS. Therefore, as the velocity increases, users' requests will leave the FBS queue sooner and the MQL will decrease. For instance, when the arrival rate σ is equal to 6000, the MQL is very close to queue capacity at a velocity of 1 km/h. But when the mobile FBS starts moving faster at speed of 60 km/h, the *MQL* is equal to four requests.

In Figure 5.5, throughput of the system is presented as a function of average velocity of the mobile FBS for different values of arrival rates. The parameters are same as the parameters used in Figure 5.4. It is obvious that as arrival rate increases, more requests are served and throughput will increase too. It has been also observed that as the mobile FBS moves faster, the smart grid throughput decreases. This is because when the velocity increases and the mobile FBS moves faster, the femtocell users are removed away from the FBS before they are served. Therefore, the number of served requests decreases, and consequently, throughput of the grid decreases as well. It's worth pointing out that for both Figures 5.4 and 5.5, simulation results are also performed comparatively for validation purposes. The maximum discrepancy between the analytical results and simulation is 1.96% and 0.07% for Figures 5.4 and 5.5, respectively, which is less than the confidence interval 5%.

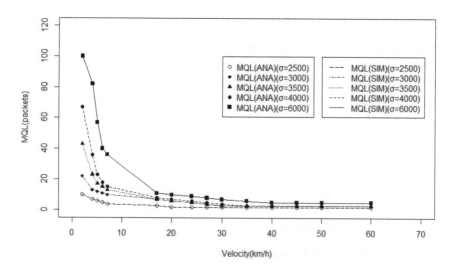

FIGURE 5.4   The effect of velocity of mobile FBSs on MQL.

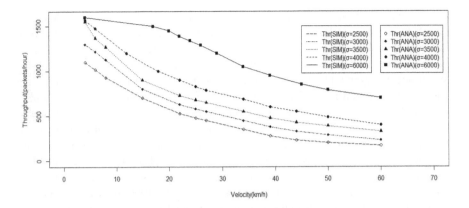

FIGURE 5.5 The effect of velocity of mobile FBSs on throughput.

5.5.2.2 The Impact of Traffic Load on Energy Consumption

Generally, when $d \leq R$, the single-hop communication is considered to be more energy efficient. Particularly, because of the low-path-loss exponents and because the distance for one hop is close to the perfect value of SNR, the start-up power overhead makes the multi-hop strategy inefficient for hop distances less than d. The assumed reduction in energy consumption is considered a central issue and should be utilized in terms of reducing the power overhead through a new strategy for HetNets where FBS can be deployed everywhere to provide a single-hop communication in connecting heterogeneous nodes, namely, sensors, PDAs, tablets, etc. And thus, multiple channel FBS can dramatically decrease power consumption while relying on single-hop communications. Figure 5.6 represents the amount of energy consumed while MQL is increasing. Obviously, the increasing MQL parameter here causes undesired

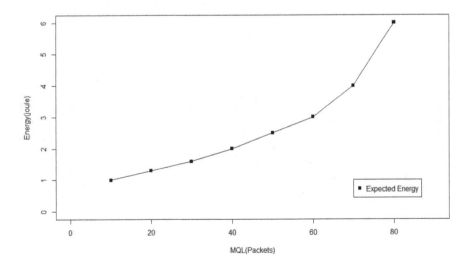

FIGURE 5.6 Energy spent per hour vs. the average MQL.

exponential energy increase. And thus, more attention shall be given for the queue length in HetNets. In Figure 5.7, the effect of how much dense is the environment in terms of mobile FBS users is studied against the average energy consumption, as well. The service rate μ is used to reflect the traffic load in the vicinity of an FBS. We notice that even with relatively large increments in user/arrivals counts, the FBS energy consumption is not that much affected. Unlike the effect of MQL, number of served users can barely affect the consumed FBS energy, where the increment in energy is linear instead of the observed exponential one shown in Figure 5.6. This can be returned to the mobility factor, where a great portion of these arrivals are leaving the queue due to varying communication range conditions (d shall be much less than R, which is the macrocell range). Nevertheless, as the service rate increases, the more energy is consumed of course. For example, at $\mu = 1000$, the expected energy consumption is 5.3 *joules*, which is ten times higher when μ is equal to 100.

In Figure 5.8, we show the effect of the service rate against the response time (R). We note that as the service rate increases, the response time starts to decrease significantly. For instance, at $\mu = 1000$, response time is 5.4 seconds which is 60 times lower when $\mu = 100$ users/h. By merging Figures 5.7 and 5.8 into one plot, we can examine the optimal service rate that guarantees the fastest response time against the lowest energy consumption. As it is shown in Figure 5.9, the optimum area is obtained at $\mu = 380$ users/h at which the response time is equal to 1.8 seconds and the expected energy consumption is equal to 1.59 *joules*. In addition, Figure 5.9 shows that there is a trade-off between energy consumption and the performance of the FBS as expected. It also shows that the proposed Q-model can play a key role in specifying the operative space and performance level, as well as energy consumption in smart grid systems.

Figure 5.10 depicts the energy consumption of the macrocell, with varying arrival rates in relation to the number of FBSs per cell. The energy-saving trend in smart grids reflects that there is some power saving achieved through the deployment of

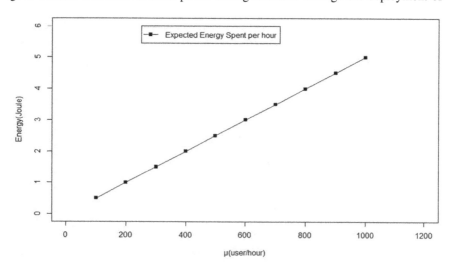

FIGURE 5.7   Energy spent per hour vs. the average service rate μ.

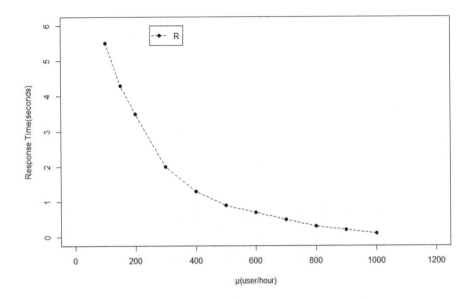

FIGURE 5.8   Response time vs. the service rate.

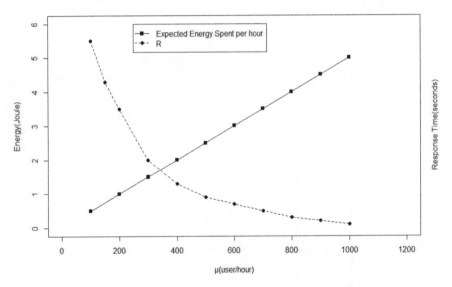

FIGURE 5.9   Response time and energy spent per hour as a function of the service rate.

FBSs while experiencing high and low arrival rates (i.e., $\sigma = 2500$ and $\sigma = 6000$). The trend reflects that the energy saving increases from 6.9% to about 9.45% when the number of FBSs per cell increases from 10 to 20 when $\sigma = 6000$. The power-saving trends for $\sigma = 2000$ reflect that quite smaller power saving is achieved through the introduction of the FBS. The lower power saving achieved while experiencing $\sigma = 2000$ can be explained by the efficient energy consumption model proposed

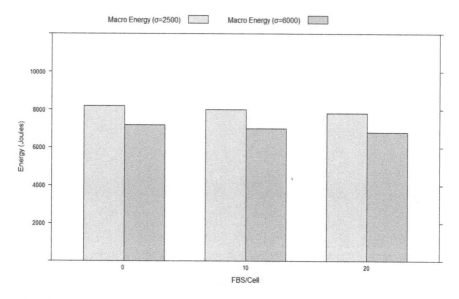

FIGURE 5.10   Energy consumption of the macrocell as a function of FBS per cell while the arrival rate is varying.

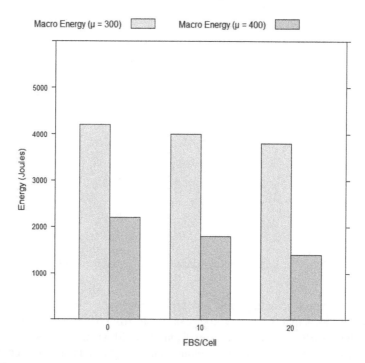

FIGURE 5.11   Energy consumption of the macrocell as a function of FBS per cell while the service rate is varying.

in this research in predicting and tolerating the extra incoming load and reacting accordingly.

In order to investigate more the increasing energy-saving trend while increasing the count of FBSs, the energy saving is evaluated for two different service rates: $\mu = 300$ and 400 users/h. The results are depicted in Figure 5.11. It can be seen that the power saving is doubled for both $\mu = 300$ and $\mu = 400$. Therefore, the increasing number of FBSs is contributing more towards the power saving when the service rate value is closer to the optimum value. Thus, we can conclude from Figures 5.10 and 5.11 that the FBS deployment in smart grids with heavy traffic loads can be an energy-efficient solution if the optimal service rate value is used. In Figure 5.11, although the total number of FBSs is the same, but the energy saving is quite better while experiencing a closer service rate μ to the optimized value that has been shown in Figure 5.9. Thus, the optimal number of FBS depends significantly on the usage of the optimal μ value.

5.6 CONCLUSION

In this chapter, a hybrid HetNet consisting of a macrocell and several mobile FBSs is considered for a smart grid application in the presence of cut-offs and energy-saving assumptions. With the introduction of Internet of Things, we believe that the aid of highly available femtocells in such setups for the increasing traffic load will become very important. A new system model is presented for mobile FBSs which are commonly deployed in macrocells to support extreme data traffic in green smart grid applications. Achieved results show that traffic load and velocity of mobile FBSs are crucial parameters affecting the mean energy consumed by the FBS. Increasing traffic load in the femtocell leads to increases in MQL, throughput, and delay. Moreover, achieved results show that for medium- and high-velocity mobile FBSs, traffic load does not affect the response time as significantly as the low mobility environment. The FBS deployment in smart grids with heavy traffic loads can be an energy-efficient solution if the optimal service rate value is considered. The optimal number of FBS depends also on the average service rate value. Meanwhile, the proposed model can provide a threshold for the mean energy consumption while expecting a specific response time. Such a model can be quite useful in specifying the operative space for FBSs. In general, this model can have a key impact on the next generation HetNets design and planning aspects.

REFERENCES

1. CISCO, "Cisco visual networking index (VNI): VNI mobile forecast highlights", 2015–2020, http://www.cisco.com/assets/sol/sp/vni/forecast_highlights_mobile/index.html, 2016.
2. G. Singh, and F. Al-Turjman, "Learning Data Delivery Paths in QoI-Aware Information-Centric Sensor Networks", *IEEE Internet of Things Journal*, vol. 3, no. 4, pp. 572–580, 2016.
3. M. Morte, "E-mobility and multiagent systems in smart grid", In *Proceeding of the IEEE International Conference on Electric Power Engineering* (EPE), Karlsruhe, Germany, pp. 1–4, 2016.

4. A. Banote, V. Ubale, and G. Khaire, "Energy Efficient Communication Using Femtocell-A Review", *International Journal of Electronics, Communication & Instrumentation Engineering Research and Development*, vol. 3, no. 1, pp. 229–236, 2013.
5. E. Bou-Harb, et. al., "Communication Security for Smart Grid Distribution Networks", *IEEE Communications Magazine*, vol. 51, no. 1, pp. 42–49, 2013.
6. M. Qutqut, et. al., "Dynamic small cell placement strategies for LTE heterogeneous networks", In *Proceeding of the IEEE Symposium on Computers and Communications (ISCC)*, Madeira, Portugal, 2014, pp. 1–6.
7. M. Qutqut, et. al., "MFW: Mobile femto-cells utilizing WiFi", In *Proceeding of the IEEE International Conference on Communications (ICC)*, Budapest, Hungary, 2013, pp. 5020–5024.
8. M. Z. Hasan, et. al., "A Survey on Multipath Routing Protocols for QoS Assurances in Real-Time Multimedia Wireless Sensor Networks", *IEEE Communications Surveys and Tutorials*, 2017. DOI: 10.1109/COMST.2017.2661201.
9. A. Law, "Statistical analysis of simulation output data: The practical state of the art", In *Proceedings of the IEEE Simulation Conference*, Washington, DC, USA, 2007, pp. 77–83.
10. M. Arshad, A. Vastberg, and T. Edler, "Energy efficiency gains through traffic offloading and traffic expansion in joint Macro Pico deployment", In *Proceedings of the IEEE International Conference on Wireless Communications and Networking (WCNC)*, Paris, France, 2012, pp. 2203–2208.
11. R. Riggio, and D. Leith, "A measurement-based model of energy consumption in femtocells", In *Proceedings of the IEEE International Conference on Wireless Days*, Dublin, Ireland, 2012, pp. 1–5.
12. M. Qutqut, et al., "HOF: A history-based offloading framework for LTE networks using mobile small cells and WiFi", In *Proceedings of the IEEE Local Computer Networks (LCN)*, Sydney, Australia, 2013, pp. 77–83.
13. M. Marsan, and M. Meo, "Queueing Systems to Study the Energy Consumption of a Campus WLAN", *Computer Networks*, vol. 66, no. 1, pp. 82–93, 2014.
14. A. Silva, M. Meo, and M. Marsan, "Energy-Performance Trade-Off in Dense WLANs: A Queuing Study", *Computer Networks*, vol. 56, no. 1, pp. 2522–2537, 2012.
15. F. Al-Turjman, "Modelling Green Femtocells in Smart-grids," *Springer Mobile Networks and Applications*, vol. 23, no. 4, pp. 940–955, 2018.
16. F. Al-Turjman, "Information-Centric Framework for the Internet of Things (IoT): Traffic Modelling & Optimization," *Elsevier Future Generation Computer Systems*, vol. 80, no. 1, pp. 63–75, 2017.
17. T. Elkourdi, and O. Simeone, "Femtocell as a Relay: An Outage Analysis", *IEEE Transactions on Wireless Communications*, vol. 10, no. 12, pp. 4204–4213, 2011.
18. F. Al-Turjman, "Information-Centric Sensor Networks for Cognitive IoT: An Overview", *Annals of Telecommunications*, vol. 72, no. 1, pp. 1–16, 2016.
19. J. Gong, S. Zhou, and Z. Niu, "Queuing on energy-efficient wireless transmissions with adaptive modulation and coding", In *Proceedings of the IEEE International Conference on Communications (ICC)*, Kyoto, Japan, 2011, pp. 1–5.
20. V. Borodakiy, et al., "Modelling and performance analysis of pre-emption based radio admission control scheme for video conferencing over LTE", In *Proceedings of the ITU Kaleidoscope Academic Conference*, St. Petersburg, Russian Federation, 2014, pp. 53–59.
21. M. Chowdhury, et al., "Service quality improvement of mobile users in vehicular environment by mobile femtocell network deployment", In *Proceedings of the International Conference on ICT Convergence (ICTC)*, Seoul, Korea, 2011, pp. 194–198.
22. O. Karimi, J. Liu, and C. Wang, "Seamless Wireless Connectivity for Multimedia Services in High Speed Trains", *IEEE Selected Areas on Communications*, vol. 30, no. 4, pp. 729–739, 2012.

23. F. Saghezchi, A. Radwan, and J. Rodriguez, "Energy-Aware Relay Selection in Cooperative Wireless Networks: An Assignment Game Approach", *Ad Hoc Networks*, vol. 56, no. 1, pp. 96–108, 2017.
24. R. Baloch, et al., "A Mathematical Model for Wireless Channel Allocation and Handoff Schemes", *Telecommunication Systems*, vol. 45, no. 4, 2010, pp. 275–287.
25. Q. Zeng, and D. Agrawal, "Modeling and Efficient Handling of Handoffs in Integrated Wireless Mobile Networks", *IEEE Transactions on Vehicular Technology*, vol. 51, no. 6, pp. 1469–1478, 2002.
26. V. Sucasas, "A Survey on Clustering Techniques for Cooperative Wireless Networks", *Ad Hoc Networks*, vol. 47, no. 1, pp. 53–81, 2016.
27. A. Mukherjee, et al., "Femtocell Based Green Power Consumption Methods for Mobile Network", *Computer Networks*, vol. 57, no. 1, pp. 162–178, 2013.
28. K. Trivedi, and S. Dharmaraja, X. Ma, "Analytic Modeling of Handoffs in Wireless Cellular Networks", *Information Sciences,* vol. 148, pp. 155–166, 2002.
29. I. El Bouabidi, et. al., "Design and Analysis of Secure Host-Based Mobility Protocol for Wireless Heterogeneous Networks", *Journal of Supercomputing,* vol. 70, no. 1, pp. 1036–1050, 2014.
30. H. Beigy, and M. Meybodi, "A Learning Automata-Based Adaptive Uniform Fractional Guard Channel Algorithm", *Journal of Supercomputing*, vol. 71, no. 1, pp. 871–893, 2015.
31. F. Al-Turjman, and H. Hassanein, "Towards Augmented Connectivity with Delay Constraints in WSN Federation", *International Journal of Ad Hoc and Ubiquitous Computing*, vol. 11, no. 2, pp. 97–108, 2012.
32. M. Deruyck, et. al., "Modelling the power consumption in femtocell networks", In *Proceeding of the IEEE Wireless Communications and Networking Conference*, Paris, France, 2012, pp. 30–35.
33. W. Wang, G. Shen, "Energy efficiency of heterogeneous cellular network", In *Proceedings of the IEEE Vehicular Technology Conference*, Taipei, Taiwan, 2010, pp. 1–5.
34. R. Chakka, "Spectral Expansion Solution for Some Finite Capacity Queues", *Annals of Operations Research,* vol. 79, pp. 27–44, 1998.
35. F. Al-Turjman, and M. Imran, "Energy Efficiency Perspectives of Femtocells in Internet of Things: Recent Advances and Challenges," *IEEE Access Journal*, vol. 5, pp. 26808–26818, 2017.
36. A. Al-Fagih, et al., "A Priced Public Sensing Framework for Heterogeneous IoT Architectures", *IEEE Transactions on Emerging Topics in Computing,* vol. 1, no. 1, pp. 133–147, 2013.
37. G. Singh, and F. Al-Turjman, "A Data Delivery Framework for Cognitive Information-centric Sensor Networks in Smart Outdoor Monitoring", *Computer Communications,* vol. 74, no. 1, pp. 38–51, 2016.
38. G. Solmaz, M. Akbas, and D. Turgut, "A Mobility Model of Theme Park Visitors", *IEEE Transactions on Mobile Computing*, vol. 14, no. 12, pp. 2406–2418, 2015.

6 Mobile Couriers and the Grid

Fadi Al-Turjman
Near East University

CONTENTS

6.1 INTRODUCTION

Wireless sensor network (WSN) has come a long way, from their support in area-specific deployments such as irrigation systems, health care, and supply chains, to supporting multiuser systems that enable simultaneous access of application that operate in large-scale Internet of Things (IoT) paradigm [1,2]. Smart cities are examples of such applications that support multiuser access on a multi-application platform. These users may want to access information such as the availability of parking space in the city, electric grid information from the smart meters around the city such as electricity consumption levels and peak hours, and/or major road accidents and any other reported emergencies. Smart cities are expected to have a grid of WSNs to provide access to large-scale information [3]. In smart cities, different challenges related to type/nature of surrounding buildings and obstacles' need to be taken into account. In this study, cities with high-rise buildings are assumed, which can severely attenuate the signal strength/quality, and accordingly, users may experience poor service. In addition, varying distances between the base station and users due to mobility is assumed. This can also be a key challenge in the assumed smart city setup where fixed transmission range from the base station can lead to significant waste in energy in case it is set to the maximum unnecessary coverage distance. Additionally, in dense deployment of wireless user equipment, interference may happen and degrades quality of service (QoS) and bandwidth utilization in homogenous networks functioning over unique frequencies.

Pervasive sensing (PS) is one great example that uses low-cost sensory devices in mobile devices to create a large-scale network for transferring data among users for the greater good of the public in smart cities [2,4]. The proliferation of wireless sensors has given rise to PS as a vibrant data-sharing model. This vision can be extended under the umbrella of the IoT to include versatile data sources within smart cities such as cell phones; radio frequency identification tags; and sensors on roads, beaches, and living spaces. The facilitation of such a vision faces many challenges under the outdoor harsh operational conditions in terms of interoperability, resource management, and pricing. With PS incorporated into IoT, it will be able to extend to data generating/sharing systems including wireless networks (WNs), IoT operating systems, database centers, and personal and environmental monitoring devices deployed in both cities and urban regions. Smart city projects in the urban regions introduce innovation in the provision of infrastructure services making applications accessible directly and in large scale. Smart grid initiatives for utilities represent part of this big scenario of smart cities. The innovation of the smart grid can affect almost all services necessary for the economic development and the well-fare state of human beings in smart cities [5–7]. Smart grid has been evolved recently in managing our vast electricity demands in a sustainable, smart, and economic manner, while utilizing already existing heterogeneous networks' (HetNets) infrastructures. The smart grid is simply an energy network that can automatically monitor the flow of electricity in a city and adjust to changes in users' demands accordingly. It comes with smart meters which are connected to the Internet to provide consumers/suppliers with smart decisions on their on-going energy usage/production. For example, a number of smart home appliances, viz., dishwashers, washing machines, and air conditioners can communicate with the grid using these smart meters and automatically manage their electricity usage to avoid peak times and make more profit. Moreover, these smart meters can be used in "smart parking lots" projects where arrival and departure times of various vehicles are traced and measured all over the city. Therefore, these parking lots have to be planned in a way that takes the average number of cars in every region into account. This service is applicable based on sensors deployed on the roads of the city and intelligent displays which inform drivers about the best place for parking in the nearby region while being synchronized with these smart meters. In this way, drivers can find a cost-effective place for parking faster, and reduce CO_2 emissions and traffic congestions in the city.

Within a smart and green grid, mobility has to be intended as the way in which customers can access and explore the grid resources using advanced and eco-friendly electricity modes. This implies being aware of the available energy resources and their real values (in term of cost, time, and carbon emissions), as well as a simple and unified access to mobile carriers in a PS paradigm. These mobile carriers can connect the different system entities, including mobile users', power lines, parking lots, and appliances at home by simply establishing a dynamic infrastructure between the distributed smart meters and their base stations in a PS paradigm. PS in smart cities can significantly affect citizens' satisfaction levels and can be realized by combining private and public networks, including the IoT as the main platform.

In order to support smart grid applications, utility companies have widely adopted wireless mesh topologies that aim to be open and interoperable HetNets. However, these HetNets solutions have yet lots of work to be done. For IoT services including smart grids, there are genetic-based meta-heuristic solutions including the effective fitness (EF) and hybrid search (HS) algorithms [8,9]. EF considers the motion direction of the mobile carrier with some environmental effects, such as surrounding obstacles and available paths, whereas HS is a modified effective method that takes into consideration speed of the motion to compute the fitness value. On the other hand, a non-genetic up-to-date approach, called LoRaWAN [10], has been recently proposed to address data delivery problem between the smart grid meters. LoRaWAN represents a convergent alternative which also adopts open and interoperable standards, but at a lower cost if compared to existing mesh HetNets including WiMAX, LTE, Wi-Fi, Li-Fi, and point-to-multipoint radio systems.

In this chapter, we change the meaning of the term "sensor" to include any data source device that is moving or stationary. Hence, this huge network created by these sensors provides a multitude of services that improve the standards and quality of living in smart cities. There are abundant sensor devices in such a setting, onboard private or public vehicles and\or deployed on roads or buildings. Such a comprehensive PS system introduces challenging implications in practice regarding the sensory system's limitations in terms of energy consumption, available bandwidth capacity, and delay [11,12]. Moreover, management is also put to test, as the data transferred across the system is of wide variety. Such a system   with multiple applications and multiple sensors on the same platform is depicted in Figure 6.1a. In this model of a smart city, there are a number of access points (APs), which communicate with sensor nodes (SNs) (or smart meters), users, and each other, to collect or gather requested information and send it to data sink (or base station or BS) based on service requirement.

In a smart city, it is common to sometimes have more than one BS collaborating together. We assume this setting in this research in contrast to other related works in the literature [13,14]. In smart cities, we exercise the ability to handle multiple users with different attributes such as latency, reliability, and throughput simultaneously. This is one important area that has not received sufficient research attention yet. The complexity involved in handling the heterogeneous traffic flows in the underlying sensor networks comes from simultaneous user request with diverse requirement. To curb the complexity problem, we propose a method that utilizes mobile smart devices called data couriers (DCs). To reduce the total number of DCs and their collaborative travel distance, we propose an approach called Hybrid

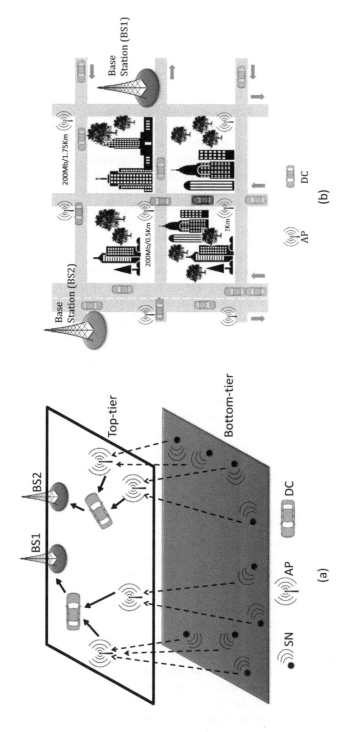

FIGURE 6.1 (a) Two-tier PS architecture, and (b) selected DCs for data collecting in smart cities.

Collaborative Path Finder (HCPF). The targeted problem can be stated as follows: Given a set of DCs with a limited storage and predetermined trajectories under the coverage of a set of APs, find the minimum DCs count that can deliver these APs' data traffic while retaining storage capacity constraints and minimum traveled distances by DCs.

To solve the above-stated multi-objective problem, we use a hybrid meta-heuristic path finder approach, based on pure genetic and local search plus (LS+) algorithms [15]. We summarize the main contributions in this chapter towards solving the problem as follows:

1. We propose a framework for PS in smart cities based on an IoT architectural model that integrates heterogeneous networks and data sources to support the smart grid project.
2. We propose a massive data collecting network for DCs in smart cities. We do this by utilizing moving sensors that collect data in a smart environment efficiently in terms of the number of DCs involved and the total distance traveled by each DC.
3. For DCs that operate in competitive time complexities and experience traffic/memory capacity constraints, we propose a collaborative genetic-based approach, called HCPF.
4. We also propose a cost-based fitness function for DC election in collaborative PS paradigm. In this function, we consider DC resource limitation in terms of its DC count, storage capacity, energy consumption, and communication link quality.
5. We provide a dynamic two-tier scheme that adheres to the social welfare of the PS system by incorporating lifetime and capacity constraints while considering, delay, and quality metrics to assure the maximum gain.
6. We perform extensive comparisons between HCPF approach and other heuristics (e.g., EF and HS), and accordingly we make significant recommendations regarding such kind of heuristic approaches in smart cities.

The remainder of this chapter has been organized as follows. Section 6.2 reviews related work in the literature. Section 6.3 provides the utilized system models in this study. In Section 6.4, we provide the details of the proposed HCPF approach. Next, we elaborate more on the HCPF via a use case in Section 6.5. The proposed HCPF approach is validated in Section 6.6 via extensive simulation/experimental results. Finally, our work is concluded in Section 6.7.

6.2 RELATED WORK

The recent explosion of mobile devices in PS paradigm, viz., smartphones, tablets, and on board sensors, inspired a category of smart grid prototypes. These prototypes can use sensor-enabled mobile devices/vehicles to monitor their local environments via smart meters, their private spaces (e.g., monitor bodily vital signals) or create a binding between tasks and the physical world (e.g., take video or audio samples). A good example for such a PS paradigm is the MetroSense presented in

Ref. [16] as a wider vision of a people-centric paradigm for urban sensing. It solely explores sensor-embedded mobile phones to support personal and public sensing. PS can be more effective while considering plenty of cloud services as well. Joining the capabilities of a multitude of smart devices in a Cloud of Things (CoT) has been proposed in several works [17–19]. Proposals specifically address smart city applications can be found in Ref. [20–22]. In Ref. [20], for instance, the authors highlighted how future cities need to collect data from an abundance of low-cost urban sensors including environmental sensors, GPS devices, and building sensors. The key idea for getting high-quality services from such cheap sensors is the cross-correlation of sensed data from several sensors and their analysis with sophisticated algorithms.

The application of PS in smart cities has recently proven to be very effective with the IoT evolution [4,23]. In a paradigm such as the IoT, data collection is carried out by deploying sensors around the city, which periodically send data about the smart city variables (e.g., temperature, humidity, and traffic conditions), to the processing center (BS) through a wireless link. To correctly collect data about a given sensor, it is advised to use a number of mobile DCs and effectively utilize their path of interest. These DCs are usually powered by low-power batteries and mounted on moving vehicles such as taxis and public transport means. Hence, the DCs in this framework will have low data transmission power, low memory capacity, and low processing power, which consequently leads to a major problem faced by most reliable smart city grid where the main challenge is connecting with the main BS. Additionally, mobile DCs must be used due to the monitoring of non-stationary phenomenon such as animals, cars, and people and the very limited network resources, such as communication range and energy budget, when it comes to large-scale applications like the ones found in smart cities. Mobile DCs were proposed in Ref. [24], given that centralized knowledge and decisions are made at the BS. Mobile SNs move along a predefined path in the sensing environment. There is a proven benefit of using DCs (mobile relays) over conventional WSN using static sensor nodes (SNs). A network that uses the former nodes has more lifetime than a network that uses the latter ones. The authors in Refs. [25,26] state that DCs were first used to prolong the WSN lifetime. The lifetime of the network is divided into equal parts called rounds. Based on a centralized algorithm running at the BS, DCs are placed at the beginning of each round. The main goal was to minimize the average amount of energy used during one round. It was concluded that the optimal locations according to this objective function remain optimal, even after the objective is changed to minimizing the maximum energy consumed per SN. However, using this energy metric to find the optimal position of mobile DCs is not the most effective means, because the solution will not be in terms of time. Therefore, the DC location found might be far from the optimal positions in a smart city setup.

Nevertheless, none of the aforementioned approaches introduce a comprehensive framework that addresses delivery, resource management, and cost challenges together. Most importantly and from the delivery perspective, they either apply simplistic communication protocols that are either basically cellular (e.g., GPRS) coupled with some Wi-Fi or Bluetooth capabilities, or apply algorithms that are intended originally for mobile Ad-hoc networks such as Ad-hoc On demand Distance Vector Routing (AODV) [27] and dynamic source routing (DSR) [28] which are short-path routing protocols based on minimum-hop count only. Genetic algorithm (GA) does not break easily and is resistant towards noise unlike the mobile DC approaches mentioned earlier in [15,29]. Each chromosome contains information, and the population means a big data for the current generation. We can use GA to simultaneously search for candidate solutions hence it can be seen as a parallel searching approach. There are numerous benefits that come with the fact that GA is a powerful and robust optimization technique. The applicability of GA, as an artificial intelligence technique, comes from the fittest chromosomes along/over consecutive generations for solving a problem. With the progressive generation of solutions, chromosomes with potential solutions become more adaptable to the environment. As a result, successful results are obtained. According to Refs. [8,9,30,31], GA has a potential for solving problems such path planning and data gathering. The authors in Ref. [30] use the GA approach to solve the problem of point-to-point path for an offline autonomous mobile robot. The GA approach was used to effectively find the optimal path with reduced cost and computation time and smaller number of generations in a static field with known obstacles. The authors in Ref. [31] use GA to calculate an optimal path for autonomous mobile robots, where a robot is represented by its coaxial coordinates in its present environment. GA uses random searches, however it actually directs searches into regions of better performance in the inside the search area [8,9]. Since the genes from good chromosomes go through the population, they are expected to handle better offsprings than the parent. In this research, we have mainly focused on genetic-based meta-heuristic approaches that consist of EF and Hybrid Search (HS) algorithms. For comprehensive analysis, another non-genetic up-to-date approach, called LoRaWAN [10], that best fit big scenarios such as the PS in smart cities has been considered. LoRaWAN is a fully convergent technology based on open standards, low cost, and was designed from the start to build urban platforms for IoT. It is best suited to the big scenario of a smart city. Unlike other Mesh topology-based approaches, LoRaWAN has a star topology which simplifies the operation, and significantly reduces traffic on network destined to routing information. On the other hand, it does not count with the possibility of coverage extension through the relay on neighboring terminal device. In Ref. [8], EF considers the motion direction with some environmental effects to develop an EF function to be used in the GA. In Ref. [9], the authors proposed a modified effective method, called HS, that takes into consideration speed of the motion to compute the composite fitness value. HS and EF are seeking the maximum fitness value for their designed fitness function, whereas others are seeking the minimum fitness which makes them less efficient in terms of the archived solution optimality. Accordingly, we focused on genetic-based approaches which are efficient in terms of time complexity and optimality as well as practicality in solving the aforementioned path planning problem. Unfortunately, some parameters such as the total traveled distance and the current application circumstances were not considered in the aforementioned collaborative approaches. Therefore, we propose a hybrid meta-heuristic approach which employs a modified version of the local search (LS) [32] in collaborative manner to improve the path search in smart cities based on application-specific characteristics, such as delay sensitivity and bandwidth availability.

6.3 SYSTEM MODELS

In this section, we discuss the main system models we assume in our HCPF approach. We start with the assumed notations, followed by the network, energy, and communication models which are considered in this study.

6.3.1 NOTATIONS

The following is the list of the assumed GA components mapped to the aforementioned DC election problem:

- *Gene* is represented by a decimal value that can be between zero and the number of APs.
- *Chromosome* is a candidate solution to the given problem and is composed of several genes. It is also called individual and coded as a finite length vector of genes (represented in bit strings).
- *Search space* is the area in which the search is performed to find the solution among the chromosomes.
- *Fitness value* is the associated with each chromosome, and it represents quality of the candidate solution.
- *Population* consists of a number of chromosomes and is maintained within a specific search space.
- *Generation* is the current population in any stage of the whole searching process.
- *Genetic operator* is a variety of operations to be applied on a chromosome for producing better individuals (e.g., crossing over and mutation).

GA combines and passes the information of an individual to any other individual in order to produce new solutions with good information inherited from previously found solutions. It is expected that these new individuals, called offspring, are hopefully leading towards optimality; and thus, the new population can be better than the previous one over successive generations towards the global optimum. This process continues until some condition (termination criteria), i.e., the maximum number of generations is reached, is satisfied.

6.3.2 NETWORK MODEL

In this chapter, we consider a multi-tier telecommunication framework with three main components: (1) the BS, (2) the data collector (DC), and (3) the WSN AP in a smart city (Figure 6.1). Telecommunication infrastructure that supports the smart grid application in IoT-based PS is divided into outdoor RF devices and indoor wired devices connected to the Internet via network interface cards (NICs). RF devices are generally installed on GSM towers, top of buildings, or even poles, and are responsible to communicate with endpoints (end users) in its covered area. The NICs are interfaces embedded on endpoints and are responsible to communicate with the nearest RF device and make interface with the Internet. Outdoor RF infrastructure is usually divided into private and public networks. Public networks are those provided by third parties, generally the telecommunication companies that share their infrastructure with several other users to provide different services such as mobile network (GPRS, 3G, 4G), dedicated circuits, satellite services, and Internet. Private are those in which a network or a partition of network is dedicated to exclusive use of a single user, and it can be own assets acquired and operated by individuals or a specific company itself. There are several alternatives which are currently available for utilities companies to construct a private telecommunication network for their implemented smart grid applications using the PS paradigm. Those alternatives can be used at the backbone layer of the network such as the optical systems, and the super-high-frequency (SHF) point-to-point radios, at the backhaul layer such as WiMAX, LTE, and point-to-multipoint radios, and/or at the access layer such as RF Mesh and LoRaWAN. Although such wide range of technologies/alternatives can add further complexity to the overall data delivery process, it provides better financial situation for the proposed PS paradigm and makes it more flexible in satisfying the different users' categories/needs in such a heterogonous system. Furthermore, heterogeneity of the system will be a promising solution to satisfy multiple users' needs in public places such as downtowns or shopping malls, where the capacity of the cellular network can be overwhelmed. In such scenarios, having Wi-Fi as a backup network would be a solution to handle the gigantic incoming data requests from the smart cities' users.

In the assumed telecommunication network, there are multiple BSs, and this can be a typical situation in smart cities, where multiple service providers can co-exist [33,34]. And thus, different DCs can move in the same route and collaborate with each other via exchanging in-city routing plans. In this chapter, a two-tier hierarchical architecture is assumed as a natural choice in large-scale applications, in addition to provide more energy-efficient solutions. The lower layer consists of SNs that sense the targeted phenomena and send measured data to the APs in the top tier, as shown in Figure 6.1a. Usually, these SNs have fixed and limited transmission ranges and do not relay traffic in order to conserve more energy. The top tier consists of APs and DCs which have better transmission range and communicate periodically with the BS to deliver the measured data in the bottom tier. APs aggregate the sensed data and coordinate the medium access, in addition to supporting DCs in relaying data from other APs to the BS in the top tier. A data packet consists of the data traffic that includes loads of a group of APs in the network. Each AP delivers its sensed data via multi-hop transmissions through other APs/DCs to a BS. DCs are equipped with wireless transceivers and are responsible for forwarding the APs' data load to the destination (AP or BS) once they are within their communication range. APs represent the target for a given DC its route. In our formulated path planning problem, we assume the following inputs. Each DC has a specific/limited storage capacity. And different APs can communicate with each other for collaboration. Moreover, we assume the following constraints for the problem:

1. Every DC is usually associated with a predetermined route/trajectory in the city.
2. Each route includes a path. The path of any route of a DC includes a subset of the APs.

3. Each smart node/APs can be assigned to more than one DC (i.e., to be visited more than once).
4. The total traffic carried by a DC cannot be more than the DC storage/BW capacity.

We assume the outputs for this problem solution consist the following three characteristics: (1) the minimum required number of DCs, (2) the total traveled distance by them, and (3) the paths of all the optimally selected DCs. It's worth pointing out here that the total traveled distance represents the consumed time and energy in the found solution, where this distance is the total driven kilometers from the data source to distention.

6.3.3 ENERGY MODEL AND LIFETIME

The energy supply of a mobile DC can be unlimited or limited. When a DC has an unconstrained energy supply (rechargeable or simply have enough energy relative to the projected lifetime of the APs), the placement of DCs is to provide connectivity to each AP with the constraint of the limited communication range of APs. When the energy supply of DCs is limited, the allocation of DCs should not only guarantee the connectivity of APs but also ensure that the paths of mobile DCs to the BS are established without violating the energy limitation. In this research, we assume a fixed and limited power supply at the DCs. In order to measure the network lifetime, a measuring unit needs to be defined. In this work, we adopt the concept of a round as the lifetime metric. A round is defined as the time period over which every irredundant SNs and relay nodes in the network communicate with the BS at least once. It can also be defined as the time span over which each EC reports to the BS at least once. At the end of every round, the total energy consumed per node can be written as

$$E_{cons}^i = \sum_{\text{Per round}} J_{tr} + \sum_{\text{Per round}} J_{rec}, \tag{6.1}$$

where $J_{tr} = L\left(\varepsilon_1 + \varepsilon_2 d^n\right)$ is the energy consumed for transmitting a data packet of length L to a receiver located d meters from the transmitter. Similarly, $J_{rc} = L\beta$ is the energy consumed for receiving a packet of the same length. In addition, ε_1, ε_1, and β are hardware-specific parameters of the used transceivers. In addition, if the initial energy E_{init} of each node is known, the remaining energy per node i at the end of each round is $E_{rem}^i = E_{init}^i - E_{cons}^i$. In order to consider the effect of packet retransmission at the MAC layer, the extended energy model in Ref. [35] has been considered. Using this model, the effect of the wireless links reliability on the energy consumption of per DC is taken into consideration. Based on Ref. [35], the total energy consumed by a DC to receive the packet is

$$E_{DC} = x \frac{P_r L_d}{R_d} + y\left(P_t + \frac{P_{DC, SN}}{k}\right) \tag{6.2}$$

where L_d (bits) denotes the size of the data packet transmitted from SN to DC, R_d denotes the data rate with which the data packet is transmitted by SN, P_t is the power required to run the processing circuit of the transmitter, P_t is the power required to run the receiving circuit, $P_{DC, SN}$ is the transmission power from SN to DC, and k be the efficiency of the power amplifier. Also, L_a (bits) denotes the size of the acknowledgment, R_a denotes the data rate by which the acknowledgment is transmitted by DC, and the values of x and y depend on reliability of the forward link between a SN and the DC for data packets and reliability of the reverse link (DC, SN) for acknowledgments.

6.3.4 COMMUNICATION MODEL

Practically, the signal level at distance from a transmitter varies depending on the surrounding environment. These variations are captured through what we call *log-normal shadowing* model. According to this model, the signal level at distance from a transmitter follows a log-normal distribution centered on the average power value at that point [36]. This can be formulated as follows:

$$P_r = K_0 - 10\gamma \log(d) - \mu d \tag{6.3}$$

where d is the Euclidian distance between the transmitter and receiver, γ is the path loss exponent calculated based on experimental data, μ is a random variable describing signal attenuation effects[1] in the monitored site, and K_0 is a constant calculated based on the transmitter, receiver and field mean heights. Let P_r equal the minimal acceptable signal level to maintain connectivity. Assume γ and K_0 in Eq. (6.3) are also known for the specific site to be monitored. Thus, a probabilistic communication model which gives the probability that two devices separated by distance d can communicate with each other are given by

$$P_c(d, \mu) = K e^{-\mu d^\gamma} \tag{6.4}$$

where $K_0 = 10 \log(K)$. Thus, the probabilistic connectivity P_c is not only a function of the distance separating the SNs but also a function of the surrounding obstacles and terrain, which can cause shadowing and multipath effects (represented by the random variable μ).

6.4 HYBRID COLLABORATIVE PATH FINDER (HCPF)

Although there has been improvement on the computing and energy resources of connected devices in smart cities, they are not enough for the unrestricted deployment of ambitious IoT systems. The mobile edge computing nowadays highlights that such resources ought to be wisely employed for shifting and distributing computing and sensing tasks towards the network edge (user). Accordingly, we based our proposed HCPF solution on opportunistic PS (PS) due to its inherent ability in sensing the surrounding environment and providing a response to detected changes (events) in timely and cost-effective manner.

PS in IoT is a capable framework to provide a convergent telecommunications infrastructure, interoperable between different manufacturers, that enables investments optimization and the provision of services at a low cost. Based on open standards, it can meet all the demands of IoT in smart cities, including the smart grid application. It can be viewed as a dynamic networked system composed of a large number of smart objects that can communicate with each other and/or with users through heterogeneous wireless connections. Mobile data collectors (DCs) are used as communication devices for the smart objects in the smart city to access the Internet and reach the end user. The average speed per DC while it moves in the smart city is assumed be between 15 and 60 km/h. By using such DCs, a significant amount of traffic can be offloaded from cellular networks and other similar global ones, which can be overloaded and inefficient during peak times of the year when dense wireless devices are served simultaneously. Therefore, the operator's cost can significantly be reduced. In addition, due to the short distance between receivers and transmitters in the proposed PS framework, the power consumption and battery of the mobile user devices can be saved and better utilized. As much as we increase the number of deployed DCs, they consume much more energy to work and this can increase the overall cost as well. Consequently, the network will not be financially affordable to be in service. And thus, the HCPF approach aims at optimizing the count of DCs through a multi-objective fitness function that takes into consideration the aforementioned smart city characteristics. This fitness function considers the available resource limitations in terms of delay, capacity, and price on the data providers' side, as well as user's quality and trust requirements from the requesters' side. In characterizing the smart city network, we use a set of locations that belong to a number of APs. We use some vehicle routing problem (VRP) instances from the literature [37] for referencing, and thus, we can know the exact deployed APs' locations. To minimize the total energy of the network, we try to minimize both the total number of DCs which is required to serve all the APs in the network and the total traveled distance by the minimized set of DCs. And now, after elaborating a bit more on the used genetic-based notations, we propose our HCPF framework to optimize the cost, calculated by DC count and traveled distance, in a smart city setup.

6.4.1   CHROMOSOME REPRESENTATION

To solve the aforementioned problem using genetic-based algorithms, we represent each feasible solution/path by one chromosome that is a chain of integers where each integer value corresponds to an AP or a BS in the network. Each DC identifier uses a separator (*zero* value) between a route pair and a string of AP identifiers to represent the sequence of deliveries on a route.

Example 1:

Assume we have nine APs and three DCs are in the city shown in Figure 6.1b. Then, in the sequence below, we give a potential solution for the problem in order to demonstrate the encoding of a chromosome representation. Note that "0" represents the BS, and the number of zeros indicates the number of required DCs in the solution.

Gene sequence: 0-5-6-3-0-7-8-2-1-6-0-8.
Path 1: BS 5 6 3
Path 2: BS 7 8 2 1 6
Path 3: BS 8

These will be the paths found for the three DCs that are used by the BS in this example.

6.4.2 GENERATING INITIAL POPULATION STRATEGY

The strategy for generating initial population for a GA is an imperative part. In this step, we choose the closest nodes for a specific node in creating the location sequence. The idea is to determine the close by nodes of each node before the creation of an initial set in the genetic pool and prevent it to be completely random. This information is accumulated over several communication rounds and leads to a knowledge base (KB) at the DC, which can be looked up by reasoning mechanisms to make quick decisions about the data delivery sub-path. This KB facilitates learning from feedback in the network about the actual values of desired QoS attributes as seen at receiver nodes. This helps in reducing the communication and energy overhead in the network. In fact, the initial population strategy doesn't wait for the establishment of an end-to-end path from source to BS. Instead, it relies on the learning from the feedback provided at each hop via Quality-aware Cognitive Routing (QCR) algorithm [38–40]. A pseudo-code description of the request dissemination and information gathering is shown in Algorithm 6.1. Steps 1 and 2 represent the beginning of stage 1, where the request is multicast to all DCs. Steps 3–5 indicate the action taken when an *exact-match* is found at a DC, at any stage of the query dissemination. Steps 6–14 indicate the actions involved in disseminating the request through nearby distributed SNs till all the information required to update a *partial-match* to *exact-match* is found. At the end of stage 1, a DC containing data that is an *exact-match* with the *Query* is identified, and the QCR algorithm is initiated. Steps 16–18 indicate the steps taken if no match was found or the request timed out. A pseudo-code description of the QCR algorithm is shown in Algorithm 6.2. Step 1 starts with acknowledging the DC with an *exact-match* to the *Query* from Algorithm 6.1. In steps 2 and 3, a next-hop routing path is chosen from the KB, based on the traffic type and QoS attribute priorities associated with it. Steps 4–11 show the actions taken at the next-hop nodes depending on whether they were SNs or DCs, and their location relative to the BS. Steps 12–15 indicate that the once data is received successfully at the receiving node, an "Ack" is sent back to the transmitter with information about the QoS of the received data.

Algorithm 6.1: Query Dissemination and Data Gathering

1. **If** BS has a new *Query* from the user
2. **Then** multicast *Query* to all DCs in the network
3. **If** an *exact-match* is found in the cache of any of the DCs

4. **Then** initiate Algorithm 6.2 (QCR) from that DC
5. **Goto** Step 18
6. **If** *partial-match* is found at a DC
7. **Then** multicast *Query* from that DC to SNs in its cluster
8. **If** *partial-match* is found at any of the SNs
9. **Then** multicast *Query* from DCs to SNs in their cluster
10. Gather *Query*-relevant data from SNs at each DC
11. **If** *exact-match* found at DC
12. **Then** transmit data from DC to the DC that issued the....*Query*
13. Aggregate the data received from SNs at the DC
14. **If** *exact-match* is verified for this data at DC
15. **Then Goto** Step 4
16. **If** no match found or request timed out
17. **Then** abort current request
18. **End**

Algorithm 6.2: QoS Aware Cognitive Routing (QCR)

1. **If** a DC with data that is an *exact-match* to *Query* is identified
2. **Then** identify the priorities with respect to QoS attributes and Energy Cost based on the attribute: "Traffictype"
3. Look-up the *KB* to identify one or more next-hop paths that satisfy the QoS requirements for the traffic type
4. Transmit data from source SN to the next-hop receiver node(s) listed in the KB
5. **If** a DC located at one-hop from the BS received the data
6. **Then** continue to relay data from that DC to the BS
7. **If** an intermediate DC received the data
8. **Then** look-up DC's routing table (RT) to identify the next-hop DC and go to step 2
9. **If** a neighboring DC received the data
10. **Then** go to step 2
11. **If** any node other than the BS received the data
12. **Then** calculate the QoS of received data and send this information back to the source SN with an "Ack"
13. The SN stores this information about the receivers and their associated QoS in its KB
14. **If** BS received the data
15. **Then** BS computes the QoS for the data received and sends it back to the transmitter(s) along with an "Ack"
16. **If** "Ack" transmitted by BS
17. **Then** data has been received for the current request
18. **End**

Lines 16–18 indicate that when the sink transmits an "Ack," it means that the requested data was successfully received and that ends the current transmission cycle. Nodes that fail to send an "Ack" in a previous transmission cycle will be chosen for up to three times without receiving an "Ack." Beyond this, the transmitter presumes the HCPF GA described in the following subsection.

6.4.3 PATH PLANNING IN HCPF

HCPF is an event-driven and collaborative-based optimization algorithm to select the best chromosome from the search space. In this section, we formulate a data path based on the available set of touring DCs in the city. We assume that each AP location is a target for DCs on its route where they stop (or satisfy speed and channel characteristics based on Eq. (6.3)) for the minimum amount of time required for exchanging data via *Wi-Fi* connections. APs are assigned to DCs based on their passing time and requested service specifications and requirement (e.g., security level and maximum delay). This service is quantified usually using a QoS factor that can be archived in collaboration with other BSs and APs. This QoS represents the successful arrival of bit-rate at the final destination (i.e., BS), where some services (e.g., VoIP and video streaming) require a higher quality level in terms of transmission rates as opposed to data/information about road-traffic updates, for instance. This is predicted in this work via log-based function running on the top level of the system (e.g., a set of accessible BSs/APs in the cloud). We formulate a multi-objective fitness function for our genetic-based approach. The fitness function $F(x)$ for the xth chromosome in our GA population aims to find solutions that takes into consideration cost, QoS, and DC motion characteristics such as speed and traveled distances. This function is formulated as follows:

$$F(x) = \lambda * R\text{Distance}(x) + \mu * R\text{Count}(x) + \beta * \text{QoS}(x) + P_c \qquad (6.5)$$

where $R\text{Distance}(x)$ computes the total traveled distance when chromosome x is applied, and the $R\text{Count}(x)$ function returns the minimum required number of DCs. This count is proportional to total cost. Furthermore, $\text{QoS}(x)$ is used to calculate the achievable bit-rate of a generated chromosome x. And P_c is the average probabilistic connectivity between the neighboring nodes on the same path, which is calculated based on Eq. (6.4). These four variables are weighted via tuning parameters (λ, μ, and β) that makes the proposed framework adaptable to the heterogeneous nodes/applications in a typical smart city's PS. Total distance is weighted by λ value equal to 0.001, and the number of DCs is weighted by μ value equal to 100 based on preliminary studies in Refs. [14] and [15]. Since total distance has more importance than the number of DCs, the coefficient of it is more sensitive, and thus, λ value is set to be 0.001. These tuning parameters formulate our fitness function that directs the search space to find chromosomes that cover all APs interested in exchanging their data loads, and then find the shortest path with the least possible

DCs. Meanwhile, for the creation of a new generation, we apply four main steps: (1) *Selection*, which is used for choosing an individual pair to apply the genetic operators (*crossover* and/or *mutation*), and in this study, we use *tournament* selection method; (2) *Crossing-Over*, which is used to add the genes of the two different parents into the child; (3) *Mutation,* which is applied for modifying some genes in the individuals, randomly selects an AP in the chromosome (test) to be flipped from one digit to another within the range of the available AP count without affecting the test logical structure; and (4) *Reproduction*, which is used for the construction of the next generation using the current parents. In this approach, we used a variant version of the *Reproduction* step, where almost every new population replaces an old one per iteration. Unlike Ref. [32], we apply an improved 2-opt LS algorithm as a *Mutation* operator. We combine our GA in Ref. [29] with the improved LS+ (applied at the end of each created generation) for better converging results. We point out here that neighborhoods of a potential solution are searched for better convergence as well.

Algorithm 6.3: Pseudo-Code of the Improved LS+ Approach.

Function: Improved_LS+()
While (local minimum is not achieved)
    Select best AP pair: $(i, i+1)$ and $(j, j+1)$
    If distance$(i, i+1)$ + distance$(j, j+1)$ > distance(i, j) + distance $(i+1, j+1)$
        Exchange the edge with 2-opt
    End If
End While

Algorithm 6.4: Pseudo-Code of the HCPF Approach.

Function: HCPF()
Inputs
    Max_generations: the maximum number of iterations to run the algorithm.
    Initial and *max_population* size.
    Crossover and *mutation* probabilities.
    The *AP sequences* in a smart-city.
Begin
    Generate initial population chromosomes from *AP sequences*.
    Compute the fitness of each chromosome using Eq. (6.5).
    Do{
    While (offspring population != parent population size)
        1st parent=Tournament_selection(2 random chromosomes)
        2nd parent=Tournament_selection(2 andom chromosomes)
        Crossover(1st parent, 2nd parent) based on *crossover probability.*
        Apply all three mutation types on the two newly-generated offspring based on *mutation probability*, using **Improved_LS+().**

Delete/Repair any logically invalid offspring.
Replace the parents by offspring if the parent has worse fitness.
Otherwise, let parent propagate to the new population.
}While (*Max_generations*)
if (population converged && population size < *max_population*)
Double the population size and add extra new randomly generated chromosomes to the population.
End
Return best solution in current generation.

In terms of time complexity analysis, we use big-O notation. This means that the time taken to solve a problem of size n can be described by $O(f(n))$, since traditional LS+ algorithm has complexity of $O(n)$ when non-enumerative (data-dependent) search is allowed [32], where n is the total count of APs in our proposed two-tier architecture. Our improved LS+, in Algorithm 6.3, has a time complexity of $O(n\log n)$. And thus, the total time complexity of the proposed HCPF approach, in Algorithm 6.4, end up to be $O(n^2\log n)$.

For verification purposes, the running time and number of populated solutions have been recorded while HCPF is running. The recorded runtime is then plotted, and polynomial fitting has been used to estimate the growth function of the running time. The theoretical value of the population growth of the proposed HCPF algorithm is evaluated as

$$P_{\max} = C_P P_0 (1 + r_c) (1 + r_m) \tag{6.6}$$

Equation (6.6) is derived from the trend of the population growth with every iteration of the HCPF algorithm. Initially, the population size is a positive value P_0. In worst case, the whole population is selected during the *selection* process. This means that P_0 number of solutions is considered for the *crossover* operation. A portion of the population is selected for the *crossover* process, which reproduces additional chromosomes/solutions to be added to the population. Let's assume r_c is the *crossover* rate where $0 \le r_c \le 1$; the number of additional chromosomes would be $P_0 r_c$. This makes the total number of chromosomes in the population so far equal to $P_0 + P_0 r_c$. by factorization, this can be simplified to

$$P_0 = P_0 (1 + r_c) (1 + r_m). \tag{6.7}$$

The expression in Eq. (6.7) is then multiplied with a coefficient of population C_P, to enable some amount of control over the maximum population. The time complexity of the HCPF algorithm is then determined from the fitted growth function to be $O(n^2\log n)$ as well. It is worth pointing out also that HCPF implementation is not straightforward as it might produce computational and energy overheads on resource-constrained devices. Accordingly, event-driven processing is applied as a natural choice to efficiently orchestrate the synchronous operation of the different stages in HCPF without incurring substantial overheads. We opted to follow an event-driven approach in which system components are only activated upon message notification for carrying out a time-bounded task. Such approach plays well with power management facilities of the employed PS paradigm in smart grid. However, as the HCPF framework heavily relies on public mobile relays, it could become vulnerable to ambient data manipulations, which brings up security issues. Moreover, considering the criticality of smart grid applications is an aspect that needs to be increasingly observed in the construction of infrastructures of telecommunication technology in order to avoid security issues. Accordingly, it is recommended to maintain the selection of the telecommunication technology to meet the requirements of high availability, and security under the control and responsibility of the company that owns/rent the infrastructure equipment. In addition, encryption methods can be applied while utilizing unregistered DCs in relying security-sensitive data. Also, privacy and anonymity issues are expected due to the personal exchanged information in the smart grid project. Therefore, it is of great relevance to keep HCPF artifacts reliable and trustworthy so as to be able to secure the core parts of it.

6.5   USE CASE

In this section, we elaborate more on our proposed HCPF approach via a typical delay-tolerant communication scenario in rural areas where people and/or public transportation can reach but the infrastructures of Wi-Fi/cellular networks are not deployed [36]. We illustrate an example where three regions in a city have the APs: *A*, *B*, and *C*, which have established one or more connecting paths, through some DCs, to the BS1 shown in Figure 6.2. Note that each DC path has its end-to-end storage capacity and traveled distance characteristics as speculated by the HCPF approach given in the format (capacity/distance) in Figure 6.2. These characteristics are based on routing table exchanges between DCs and the APs. This is achieved via

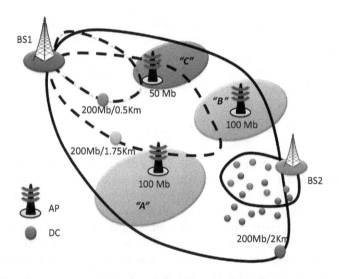

FIGURE 6.2   A use case for a group of regions' APs serving multiple BSs.

our newly developed fitness function in Eq. (6.5) which adopt these characteristics while the hybrid GA in HCPF is applied. The HCPF approach aims at selecting the paths (shown in dashed-lines) that guarantees minimum total traveled distance without violating the DC capacity constraints. The decision of utilizing these DCs depends on its resources which are stored, as well, at the AP's routing table. For example, assuming that AP in region A in Figure 6.2 has to transmit packets with a data load that requires a minimum DC capacity of 100MB, the HCPF detects two different paths connecting A to the BS. Both of them conform the capacity constraints, and the one among these two which provides the minimum total traveled distance is ultimately chosen. It will be also able to serve the AP at B as the remaining storage of that DC is still satisfying the capacity constraint. And then, the second route (dash line) passing by city C will be utilized to deliver its corresponding data. And hence, the selected paths will be

Path 1: BS1 A B
Path 2: BS1 C

Note that the non-dashed path satisfies the DC capacity constraint; however, it doesn't provide the minimum traveled distance in total if it has been chosen for A and B. The selection of the minimum DC count in the proximity of each AP in the network is achieved by periodically exchanging routing tables and/or registration records with BSs to deliver delay-tolerant data packets. This selection process is repeated at the beginning of each triggered round as we aforementioned. The mobility history of the DCs is examined against the communication range of the corresponding destination (AP). Based on the results, DC candidates are defined according to the best (i.e. minimum) traveled distance in total per round. Among the most significant processing implications is the mobility factor. Since the average experienced speed per DC in the assumed smart city setup is between 15 and 60 km/h, there will be enough time to exchange messages using the IEEE 802.15 standard. However, we classify the received signal quality into real-time (RT) and non-RT traffic for more user satisfaction, where RT users can be charged extra fees for this type of data traffic. Meanwhile, traffic load, the velocity of mobile DCs, and the multi-interface DC are all crucial parameters to achieve the best average energy consumption by the grid system. For instance, increasing the DCs' request rate in the smart city can lead to an increment in the average waiting time and energy consumption per data unit.

6.6 PERFORMANCE EVALUATION

In this section, we assess the correctness of the proposed HCPF approach first via preliminary study relying on realistic route instances experiments. In this experimental work, we compare HCPF against the pure GA and LS algorithms. Then, we evaluate the performance of the proposed HCPF against the LoRaWAN, EF, and HS algorithms in terms of varying design aspects that affect the total cost in a medium-scale city per the depicted parameters values in Tables 6.1 and 6.2. Detailed description of our experimental and simulation setup is given in the following subsections.

TABLE 6.1

Parameters of the Simulated Networks

Parameter	Value
T	70%
N_c	110
ψ	0.001 m/s
D_{max}	500 m/s
ω	200,000 km/s
γ	4.8
δ^2	10
P_r	-104 dB
K_0	42.152
R	100 m

TABLE 6.2

Components of the Simulated Networks

Simulator Component	Value
AP radius	100 m
DC velocity	Low, medium, high
AP transmit power	20 mW
Transmission bandwidth	5 MHZ
AP count	100
Expected failure rate	0.001/h

6.6.1 SIMULATION SETUP

Using MATLAB® R2016a and Simulink® 8.7, we simulate randomly generated heterogeneous networks to represent the PS environment in a smart city. A discrete event simulator is built on top of the aforementioned MATLAB platforms for more realistic results.

In this simulator, an event-based scheduling approach is taken into account which depends on the events and their effects on the system state. One of the most commonly used stopping criteria, called relative precision, is used in this simulation to be stopped at the first checkpoint when the condition $\beta < \beta_{max}$, where β_{max} is the maximum acceptable value of the relative precision of confidence intervals at $100(1 - \alpha)\%$ significance level, $0 < \beta_{max} < 1$. Accordingly, our achieved simulation results are within the confidence interval of 5% with a confidence level of 95%, where both default values for β and α are set to 0.05. Our Simulink simulator supports wireless channel temporal variations, node mobility, and node failures. The simulation lasts for 2 hours and run with the lognormal shadowing path loss model. Based on experimental measurements taken in a site of dense heterogeneous nodes [24], we adopt the described signal propagation model in Section 6.3, where we set the communication model variables as shown in Table 6.1, and μ to be a random variable that follows a log-normal distribution function with mean 0 and variance of δ^2. The parameters of the utilized network components in this study are mainly driven from Refs. [41,42] and summarized in Table 6.2.

6.6.2 EXPERIMENTAL SETUP AND BASELINE APPROACHES

We verify our proposed HCPF approach via a number of real experimental vehicular routing problem (VRP) instances taken from Ref. [37]. This step is essential to validate the correctness of the proposed genetic-based algorithm. To do this, we compare it with pure GA and LS algorithms. To use GA without LS, we set the mutation probability to 0. Also in order to use LS without GA, we set the crossing-over probability to 0. We execute our demo 100 times for each experiment and take the averages of the results. Then, we extend our assessment section via simulation-based scenarios in which we vary the aforementioned parameters (in Tables 6.1 and 6.2) for more comparative analysis about the targeted path planning problem. We assume a network of vehicles in a smart city, where DCs are a subset of the moving public transportation vehicles. We assume up to 100 total APs with one BS. For some experiments, we keep data traffic or DC capacity fixed and for other system components' values as depicted in Table 6.2. The assumed networks in this study are random in terms of their nodes' positions and densities. In order to select the most appropriate DC trajectory in these randomly generated networks, we apply our HCPF scheme. The output of the HGPF scheme is compared to output of another set of baseline approaches in the literature. These baseline approaches address the same problem tackled in this research; however, they use different path planning strategies; the EF, the HS and the long-range wide area network (LoRaWAN). In addition, we adopt three different versions of the proposed HCPF approach. These versions vary in the utilized stopping criteria of the proposed HCPF algorithm, which is the maximum number of generations. This value is set to be 200 in HCPF_1, 150 in HCPF_2, and 100 in HCPF_3.

6.6.3 PERFORMANCE METRICS AND PARAMETERS

In this research, we assess our proposed HCPF in terms of three main metrics:

1. *Throughput* is the total arrived data to the main BS via DCs per the time unit (second). It is measured in Mbps, and it reflects QoS in the system.
2. *DC count* is the number of DCs to be used for visiting all the devices in the network.
3. *Traveling distance* is the Euclidean distance between consecutive APs to be visited by a DC while traveling on the route and is measured in (km). This parameter has been chosen to reflect the consumed time and energy in every solution.

In the following, we also list the varying simulation parameters:

1. *AP count* is the number of APs in the network and represents the network size. This metric has a direct effect on DC count and total distance, and it reflects the complexity and scalability of the exploited path planning approach.
2. *DC count* is the number of DCs to be used for visiting all the devices in the network.
3. *AP load* is the amount of data to be delivered from an AP to the BS via a DC. This parameter is measured in Mbytes, which reflects the generated data traffic by these smart nodes in a smart city setup.
4. *DC capacity* is the maximum storage capacity for the utilized DCs measured in Mbytes. This parameter represents the limited hardware on the selected DC, where the sum of the collected data loads by a DC cannot be more than the capacity of that DC.
5. *Area size* represents the targeted region area measured in km². It has been chosen to assure the proposed solution efficiency in terms of scalability and different scale deployments.
6. *Solution cost* is the fitness value computed by the fitness function in Eq. (6.5) and it represents the overall cost of the achieved solution for the proposed DC allocation problem.

6.6.4 EXPERIMENTAL RESULTS

In this section, we examine the correctness of HCPF in comparison with the aforementioned GA and LS algorithms. Genetic-based parameters of these three algorithms are given in Table 6.3.

Experimental results are performed with the related location set, APs demands, and DC capacity, for each symmetrical capacitated VRP instance given in literature [37]. These instances are summarized in Table 6.4, where N is the number of APs, M is the number of DCs, and Q is the DC capacity. As shown in Table 6.4, the minimum number of DC with the minimum total distance is obtained with complete

TABLE 6.3

The Assumed Parameters' Values for Genetic Algorithms

Parameter	Approach Values		
	LS	GA	HCPF
Population size	100		
Maximum number of generations	100		
Selection method	Tournament selection		
Probability of mutation	5%	0%	5%
Probability of crossing-over	0%	80%	80%
Crossing-over operator	Permutation		
Mutation operator	2-opt LS		
Elitism	2		

TABLE 6.4

VRP Instances Used for Testing the HCPF Approach

Problem Instance	Best Known Value [37]	N	M	Q
att-n48-k4	40,002	47	4	15
A-n34-k5	778	33	5	100
An80k10	1763	79	10	100
B-n39-k5	549	38	5	100
E-n22-k4	375	21	4	6000
E-n23-k3	569	22	3	4500
E-n30-k3	534	29	3	4500
E-n51-k5	521	50	5	160
En101k8	815	100	7	200
F-n45-k4	724	44	4	2010

convergence in each run for all the instances. The average error and standard deviation values are zero which indicates stability in the proposed solutions. The results obtained are consistent as well with Refs. [43,44].

As depicted in Figures 6.3–6.5, while increasing the data load, the total traveled distances are increasing. In Figure 6.3, we are applying only global search (GA). Convergence is not sufficient in GA. In Figure 6.4, we are applying only LS. In this figure, convergence can be achieved immediately with LS, and the values are better in comparison with Figure 6.3. But using pure LS is not a good way to solve path planning problems, because LS searches the neighborhoods of the current solutions in the same pattern without any randomness in contrast to GA which leads mostly to local minima/maxima.

In Figure 6.5, we are applying the proposed meta-heuristic approach (HCPF). Saturation is obvious for our hybrid approach after a number of iterations when GA is improved by the LS algorithm. For data load test, we should state that the best

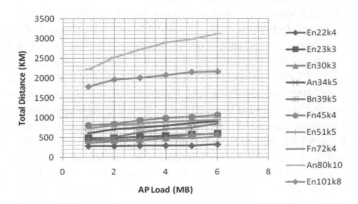

FIGURE 6.3 Data traffic vs. the total traveled distance (GA).

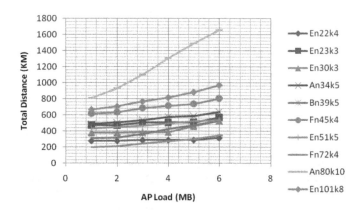

FIGURE 6.4   Data traffic vs. the total traveled distance (LS).

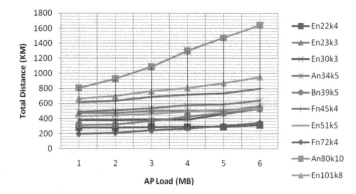

FIGURE 6.5   Data traffic vs. the total traveled distance (HCPF).

approach is the hybrid one and then LS and later GA. To observe the change in DC capacity, the data load for all APs are set to 1, and DC capacity is chosen to be 10, 20, 30, 40, 50, and 60, respectively, for all instances.

Figures 6.6–6.8 show that while increasing the DC capacity, the total traveled distances decreases. In Figure 6.6, we are applying only global search (GA). As can be seen in this figure, convergence is not sufficient for the global search without the LS algorithm. Figure 6.7 is identical with Figure 6.6, but we are applying only LS. As shown in Figure 6.7, convergence can be achieved immediately with the LS approach, and results are better in comparison with Figure 6.6. However, using pure LS algorithm without any improvement is not a good way to solve the proposed path planning problem as we have shown in Figure 6.4. In Figure 6.8, everything is identical with Figures 6.6 and 6.7, but we are applying the proposed hybrid approach (HCPF).

In Figure 6.8, the saturation is obvious for our hybrid approach after a number of iterations when the GA is improved by the LS algorithm. For the DC capacity, we should also mention that the best approach is the hybrid one and then pure LS and later GA. To observe the change in total traveled distances under the proposed approach, we vary the data load values which have been used in the literature for all the AP devices.

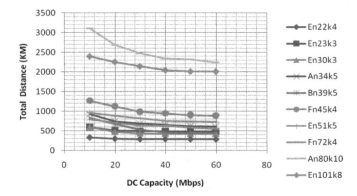

FIGURE 6.6   DC capacity vs. the total traveled distance (GA).

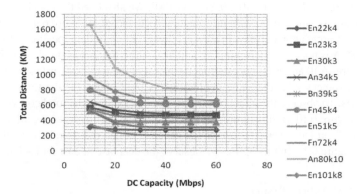

FIGURE 6.7   DC capacity vs. the total distance (LS).

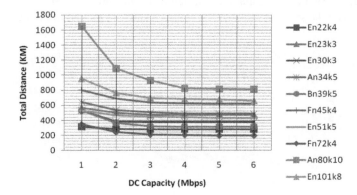

FIGURE 6.8   DC capacity vs. the total distance (HCPF).

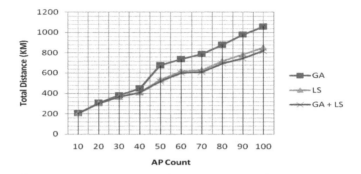

FIGURE 6.9   The three different approaches – total distance.

In Figure 6.9, we use the last instance (En101k8) in Table 6.4 which also includes DC capacity values. This instance has 100 APs, and device locations are random from the set of {10, 20, 30, 40, 50, 60, 70, 80, 90, and 100}. We also fixed the data load (traffic), the DC capacity, and the location information. As depicted in Figure 6.9, we compare the three different approaches with the same instance for random locations. It is noted that pure GA is not as good as pure LS, and also pure LS is not as good as HCPF.

In Figures 6.10 and 6.11, we examine the effect of the total cost instead of the total traveled distances with the same configurations used above for the three different approaches. To observe the effect of varying data load values, the total cost values are given in Figure 6.10. In this figure, we can observe the cost increase for all instances as the data load increases. To observe the capacity effect, the overall cost values are given in Figure 6.11. In this figure, we observe the decrease in cost for each instance.

6.6.5   SIMULATION RESULTS

In this section, simulation results are shown for various combinations of data load and DC capacity for different heuristic approaches: LoRaWAN, EF, HS, HCPF_1, HCPF_2, and HCPF_3. Simulation is started with 100 random locations and repeated 10 times with the same location set. Averaged results are plotted in Figure 6.12.

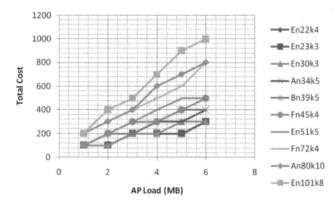

FIGURE 6.10   Cost vs. the AP data load.

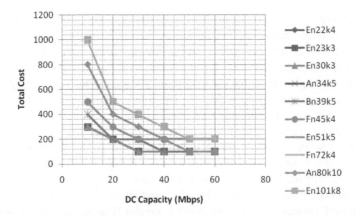

FIGURE 6.11 Cost vs. the DC capacity.

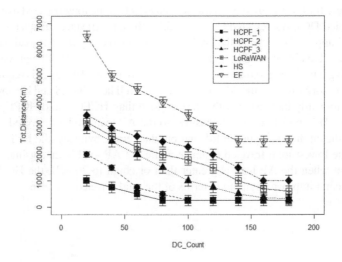

FIGURE 6.12 Traveled distance vs. the DC count.

In Figure 6.12, it is observed that total traveled distance is decreasing for all methods (i.e., the *EF*, *HS*, *LoRaWAN*, and *HCPF*), while increasing the DC counts. However, the HCPF approach outperforms LoRaWAN, EF, and HS by at least 50% of their best performance. HCPF_1 approach achieves the least total traveled distance with respect to all other baseline approaches. It's worth pointing out here that LoRaWAN outperforms typical meta-heuristic approaches when they are applied in a non-event-driven fashion such as EF and HS. Unlike these meta-heuristic approaches, the HCPF is an event-driven system that takes into consideration the different network characteristics as we aforementioned. And thus, it outperforms the LoRaWAN technique.

Moreover, HCPF_1, HCPF_2, and HCPF_3 demonstrate the same performance when the total count of DCs is greater than or equal to 150. Similarly, HCPF_1 and HCPF_2 approaches demonstrate the same performance when the DC count is greater than or equal to 100. Meanwhile, when the DC count is greater than or equal to 170, all methods cannot improve anymore in terms of the total traveled distance.

In Figure 6.13, as AP count is increasing, we notice that all approaches, except the EF-based approach, are converging to the same total traveled distance (~1800 km). It's worth pointing out here that the total traveled distance will reach an upper limit that can't be exceeded as long as the total DC count and traffic load is fixed. Unlike LoRaWAN, HS, and EF approaches, HCPF_1, HCPF_2, and HCPF_3 are monotonically increasing in terms of the total traveled distance as the DC count is increasing. This is because of the systematic steps has been followed in the proposed HCPF approach towards optimizing the search performance. We remark also that the HCPF_1 has still the lowest total traveled distance and it is the fastest approach in terms of convergence. Meanwhile, HCPF_2 and HCPF_3 approaches demonstrate the same performance when the AP count is greater than or equal to 30. While the AP count is greater than or equal to 30, all methods cannot improve anymore in terms of the total traveled distance as we remarked before.

In Figure 6.14, we study the AP traffic load effect on the total traveled distances by the occupied DCs. As depicted by Figure 6.14, the total traveled distance is increasing for all approaches, except the HCPF_1, while increasing the AP load from 10 to 80 MBytes. This indicates a great advantage for the proposed HCPF approach while using a bigger max_generation value, where no more traveled distance is required for a few more extra Mbytes at the APs. It's obvious as well how the HS and EF approaches are monotonically increasing with a larger slope than HCPF_2 and HCPF_3, which means that HS and EF are more sensitive to the AP loads to be delivered. This can be a significant drawback in the IoT and big data era. Obviously, the HCPF_1 experiences the lowest total traveled distance while the AP load is increasing. Also, we notice that when the AP load is greater than or equal to 70 MB, the HCPF_2 and HCPF_3 can't improve anymore in terms of the total traveled distance.

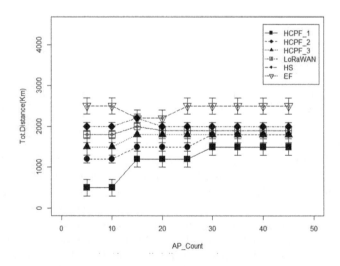

FIGURE 6.13   Traveled distance vs. the AP count.

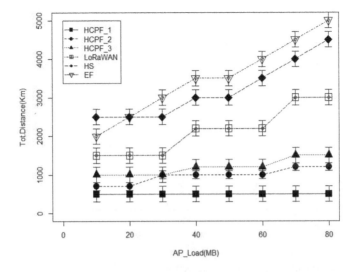

FIGURE 6.14   Traveled distance vs. the AP load.

On the other hand, the total traveled distance is decreasing for all experimented approaches while the DC capacity is increasing as shown in Figure 6.15. In this figure, HCPF_1, HCPF_2, and HCPF_3 demonstrate similar performance when the DC capacity is greater than or equal to 60. HCPF_1, HCPF_2, and HCPF_3 are very close to each other in terms of the total traveled distance as the DC capacity increases. And again HCPF_1 outperforms the other approaches in terms of the total distance. On the other hand, EF and HS are the worst in terms of the total traveled distance as the DC capacity increases.

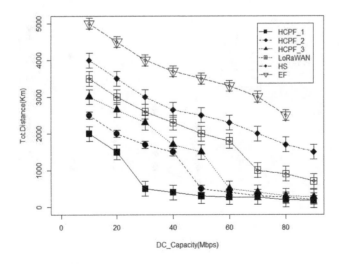

FIGURE 6.15   Traveled distance vs. the DC capacity.

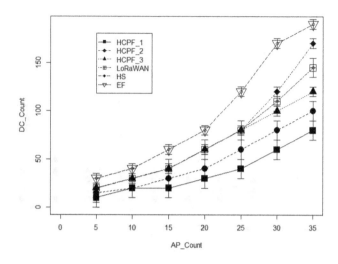

FIGURE 6.16   Required DC count vs. the AP count.

In Figure 6.16, we study the AP count effect on the total required DCs. Obviously, the DC count is increasing monotonically for all approaches while the AP count is increasing. Surprisingly, HS, LoRaWAN, and HCPF_3 approaches demonstrate the same performance when the AP count is less than or equal to 25. This can be returned for the small number of generations that has been used with the HCPF_3 approach. Meanwhile, HCPF_1 necessitates the lowest DC count while increasing the AP count. And EF necessitates the most DC count, while the AP count is increasing.

In Figure 6.17, the DC count is increasing monotonically for all methods, while the deployment area size is increasing. EF and HS approaches demonstrate the same performance when the DC count is greater than or equal to $3000\,\text{km}^2$, and

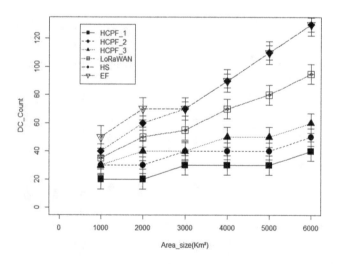

FIGURE 6.17   Required DC count vs. the targeted region size.

these approaches necessitate the largest DC count while the area size is increasing. Meanwhile, HCPF_1 necessitates the lowest DC count as the area size is increasing.

In Figure 6.18, we study the cost effect on the average system throughput measured in Mbps. As depicted in Figure 6.18, overall average system throughput is increasing for all methods, while the cost parameter is increasing. However, they saturate after reaching a specific cost value (~1000) in terms of their overall achieved throughput. For example, HCPF_3 is saturating at cost equal to 1000 and its corresponding throughput is not enhancing anymore once it reaches 127 Mbps regardless of the cost amount. HCPF_2 and HCPF_3 demonstrate the same performance once the system cost reaches a specific value (~1000). On the other hand, EF and HS approaches demonstrate almost the same performance over all cost values. However, HCPF_1 outperforms all approaches under all examined cost values. EF and HS has the lowest throughput.

In this section, we have compared total traveled distances and overall cost for the different input samples as shown in the above figures for our experimental work. As a result, we remark that the proposed hybrid approach outperforms HS and EF approaches in terms of quality and time complexity. Additionally, it's worth mentioning that under the different parameters' values and configurations, the proposed hybrid method outperforms EF or HS. However, it should be noted also that parameters and configurations have a significant impact on the achieved solutions quality and the approach convergence as elaborated in Sections 6.2 and 6.3. Another noteworthy implication of the proposed HCPF framework is the ability to simply estimate the overall framework cost for a small- to medium-scale city, where the average counts of sensing nodes per path can be controlled and assessed through Eq. (6.5). Moreover, based on Eqs. (6.5) and (6.6), QoS can be adjusted and manipulated based on the available operator/service provider budget. This can be returned of course to the emphasized edge computing principle where a considerable amount of data traffic can be offloaded from the core of the already deployed telecommunication infrastructure to those tiny casy to deploy/access SNs.

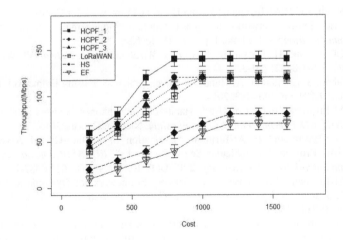

FIGURE 6.18   Throughput vs. the cost.

6.7 CONCLUSION

We introduce in this work the HCPF – a hybrid IoT PS framework for smart cities. This framework is based on a multi-tier architecture that gratifies heterogeneous data sources (e.g., sensors) with mobile data collectors in urban areas which can be isolated from the BS. According to this framework, APs at the top tier of the proposed architecture receive sensor readings and initiate delivery requests. The data delivery approach implements a genetic-based algorithm which realizes distance and cost sensitivity in competitive time complexity. At the top tier, an IoT-specific cost model uses a fitness function that maximizes the network operator gain according to the limited DCs count, storage capacity, and total traveled distances. We provide results from simulations and experimental instances in practice showing the efficiency of our framework when compared to two prominent heuristic approaches: EF and HS. Our simulation results show that the HCPF framework exhibits superior performance for different network sizes, storage capacities, DC counts, and end-to-end traveled distances. Taking into account what was studied on this work, it is recommended that utility companies that have not invested yet in a smart metering network based on genetic-based solutions should consider HCPF approach as a possible option in their projects.

Future work would investigate utilizing the same vehicles in smart city settings with non-deterministic mobility trajectories. It is also interesting to look at the same application where APs are dynamic and not only static.

NOTE

1 Wireless signals are attenuated because of shadowing and multipath effects. This refers to the fluctuation of the average received power.

REFERENCES

1. G. Singh, and F. Al-Turjman, "A Data Delivery Framework for Cognitive Information-Centric Sensor Networks in Smart Outdoor Monitoring", *Elsevier Computer Communications*, vol. 74, no. 1, pp. 38–51, 2016.
2. S. Oteafy, and H. Hassanein, *"Dynamic Wireless Sensor Networks"*, Wiley, 2014, ISBN: 978-1-84821-531-3.
3. J. Sahoo, S. Cherkaoui, A. Hafid, "A Novel Vehicular Sensing Framework for Smart Cities", LCN, pp. 490–493, 2014.
4. F. Al-Turjman, "Impact of User's Habits on Smartphones' Sensors: An Overview", *HONET-ICT International IEEE Symposium*, Nicosia, Cyprus, 2016.
5. M. Z. Hasan and F. Al-Turjman, "Evaluation of a Duty-Cycled Asynchronous X-MAC Protocol for Vehicular Sensor Networks", *EURASIP Journal on Wireless Communications and Networking*, 2017. DOI: 10.1186/s13638-017-0882-7.
6. F. Al-Turjman, "Price-based Data Delivery Framework for Dynamic and Pervasive IoT", *Elsevier Pervasive and Mobile Computing Journal*, 2017. DOI: 10.1016/j.pmcj.2017.05.001.
7. F. Al-Turjman, "Cognitive Routing Protocol for Disaster-inspired Internet of Things", *Elsevier Future Generation Computer Systems*, 2017. DOI: 10.1016/j.future.2017.03.014.

8. C. Shih, Y. Yang, M. Horng, T. Pan, and J. Pan, "An Effective Approach to Genetic Path Planning for Autonomous Underwater Glider in a Variable Ocean", In *Proceedings of International Forum on Systems and Mechatronics*, Tainan, Taiwan, 2014, pp. 1–6.

9. M. Soulignac, "Feasible and Optimal Path Planning in Strong Current Fields", *IEEE T Robot.*, 2011, DOI: 10.1109/tro.2010.2085790.

10. H. G. S. Filho, J. Pissolato Filho, and V. Moreli, "The Adequacy of LoRaWAN on Smart Grids: A Comparison with RF Mesh Technology", In *Proceedings of the IEEE Smart Cities Conference (ISC2)*, Trento, Italy, 2016.

11. D. Turgut, and L. Bölöni, "Heuristic Approaches for Transmission Scheduling in Sensor Networks with Multiple Mobile Sinks", *The Computer Journal*, vol. 54, no. 3, pp. 332–344, Oxford University Press, March 2011.

12. F. Al-Turjman, and H. Hassanein, "Towards Augmented Connectivity with Delay Constraints in WSN Federation", *International Journal of Ad Hoc and Ubiquitous Computing*, vol. 11, no. 2, pp. 97–108, 2012.

13. D. Said, S. Cherkaoui, and L. Khoukhi, "Scheduling Protocol with Load Management for EV Charging", *GLOBECOM*, pp. 362–367, 2014.

14. M. Biglarbegian, and F. Al-Turjman, "Path Planning for Data Collectors in Precision Agriculture WSNs", In *Proceedings of the International Wireless Communications and Mobile Computing Conference (IWCMC)*, Nicosia, Cyprus, pp. 483–487, 2014.

15. J. Tiu, and S. X. Yang, "Genetic Algorithm Based Path Planning for Mobile Robots", *IEEE Conference on Robotics and Automation (ICRA)*, Taipei, Taiwan, Sept. 2003.

16. A. T. Campbell, S. B. Eisenman, N. Lane, E. Miluzzo, and R. Peterson, "People-Centric Urban Sensing", In *Proceeding of the 2nd Annual International Workshop on Wireless Internet, (WICON '06)*, Boston, MA, pp. 18–31, 2006.

17. H. Patni, C. Henson, and A. Sheth, "Linked Sensor Data", In *International Symposium on Collaborative Technologies and Systems (CTS)*, Chicago, IL, pp. 362–370, 2010.

18. R. Golchay, F. L. Mouel, S. Fr'enot, and J. Ponge, "Towards Bridging IoT and Cloud Services: Proposing Smartphones as Mobile and Autonomic Service Gateways", CoRR abs/1107.4786, 2011.

19. F. Hao, T. V. Lakshman, S. Mukherjee, and H. Song, "Enhancing Dynamic Cloud-based Services Using Network Virtualization", *SIGCOMM Computer Communication Review*, vol. 40, no. 1, pp. 67–74, 2010.

20. M. Naphade, G. Banavar, C. Harrison, J. Paraszczak, and R. Morris, "Smarter Cities and Their Innovation Challenges", *IEEE Computer*, vol. 44, no. 6, pp. 32–39, Jun 2011.

21. J. Lee, S. Baik, and C. Lee, "Building an Integrated Service Management Platform for Ubiquitous Cities", *IEEE Computer*, vol. 44, no. 6, pp. 56–63, June 2011.

22. K. Su, J. Li, and H. Fu, "Smart City and the Applications", In *International Conference on Electronics, Communications and Control (ICECC)*, Ningbo, China, pp. 1028–1031, 2011.

23. F. Al-Turjman, "Cognition in Information-Centric Sensor Networks for IoT Applications: An Overview", *Annals of Telecommunications*, vol. 72, no. 3, pp. 209–219, 2017.

24. G. Solmaz, M.I. Akbas and D. Turgut, "A Mobility Model of Theme Park Visitors", *IEEE Transactions on Mobile Computing (TMC)*, vol, 14, no. 12, pp. 2406–2418, 2015.

25. F. Al-Turjman, H. Hassanein, and S. Oteafy, "Towards Augmenting Federated Wireless Sensor Networks", In *Proceedings of the IEEE International Conference on Ambient Systems, Networks and Technologies (ANT)*, Niagara Falls, ON, Canada, pp. 224–231, 2011.

26. L. Bloni, D. Turgut, S. Basagni, and C. Petrioli, "Scheduling Data Transmissions of Underwater Sensor Nodes for Maximizing Value of Information", *Proceedings of IEEE GLOBECOM'13*, pp. 460–465, Dec. 2013.

27. C. Perkins, E. Belding-Royer, and S. Das, "Ad-hoc On-Demand Distance Vector (AODV) Routing", *RFC 3561, IETF Network Working Group*, Jul 2003.
28. D. Johnson and D. Maltz, "Dynamic Source Routing in Ad-hoc Wireless Networks", *Mobile Computing*, vol. 353, ch. 5, pp. 153–181, Aug. 1996.
29. S. Al-Harbi, F. Noor, and F. Al-Turjman, "March DSS: A New Diagnostic March Test for All Memory Simple Static Faults", *IEEE Transactions on CAD of Integrated Circuits and Systems*, vol. 26, no. 9, pp. 1713–1720, Sept. 2007.
30. G. Nagib and W. Gharieb, Path planning for a mobile robot using genetic algorithms, pp. 185–189, 2004. DOI: 10.1109/ICEEC.2004.1374415.
31. M.Z. Hasan, and F. Al-Turjman, "Optimizing Multipath Routing With Guaranteed Fault Tolerance in Internet of Things," IEEE *Sensors Journal*, vol. 17, no. 19, 6463–6473, 2017.
32. C. Papadimitriou and K. Steiglitz. "On the Complexity of Local Search for the Traveling Salesman Problem," *SIAM Journal of Computing*, vol. 6, pp. 76–83, 1977. DOI: 10.1137/0206005.
33. Y. Luo, D. Turgut, and L. Boloni, "Modeling the Strategic Behavior of Drivers for Multi-Lane Highway Driving", *Journal of Intelligent Transportation Systems,* vol. 19, no. 1, pp. 45–62, 2015.
34. D. Turgut, and L. Bölöni, "Heuristic Approaches for Transmission Scheduling in Sensor Networks with Multiple Mobile Sinks", *The Computer Journal*, vol. 54, no. 3, pp. 332–344, Oxford University Press, Mar 2011.
35. T. Stephan, F. Al-Turjman, K. Joseph, B. Balusamy, and S. Srivastava, "Artificial Intelligence Inspired Energy and Spectrum Aware Cluster Based Routing Protocol for Cognitive Radio Sensor Networks," *Journal of Parallel and Distributed Computing*, vol. 142, no. 1, pp. 90–105, 2020.
36. F. Al-Turjman, "Information-Centric Framework for the Internet of Things (IoT): Traffic Modelling & Optimization", *Elsevier Future Generation Computer Systems*, vol. 80, no. 1, pp. 63–75, 2017.
37. Networking and Emerging Optimization, "Capacitated VRP Instances | Vehicle Routing Problem", http://neo.lcc.uma.es/vrp/vrp-instances/capacitated-vrp-instances/ [Accessed on February 14 2016].
38. F. Al-Turjman, "Cognitive-Node Architecture and a Deployment Strategy for the Future Sensor Networks", *Springer Mobile Networks and Applications*, 2017. DOI: 10.1007/s11036-017-0891-0.
39. F. Al-Turjman, H. Hassanein, and M. Ibnkahla, "Efficient Deployment of Wireless Sensor Networks Targeting Environment Monitoring Applications", *Elsevier: Computer Communications Journal*, vol. 36, no. 2, pp. 135–148, Jan. 2013.
40. F. Al-Turjman, "Cognitive Caching for the Future Fog Networking", *Elsevier Pervasive and Mobile Computing*, 2017. DOI: 10.1016/j.pmcj.2017.06.004.
41. J. Zhang, and G. De la Roche, et al., *Femtocells: Technologies and Deployment*, Wiley Online Library, 2010.
42. V. Chandrasekhar, J. G. Andrews, T. Muharemovic, and Z. Shen, A. "Gatherer, Power Control in Two-Tier Femtocell Networks", *IEEE Transactions on Wireless Communications,* vol. 8, no. 1, 2009, pp. 4316–4328.
43. E. Uchoa, D. Pecin, A. Pessoa, M. Poggi, A. Subramanian, and T. Vidal, "New Benchmark Instances for the Capacitated Vehicle Routing Problem", *Technical Report* - ArXiv 2014-10-4597, unpublished, 2014.
44. M. Stanojević, B. Stanojević, and M. Vujošević, "Enhanced Savings Calculation and Its Applications for Solving Capacitated Vehicle Routing Problem", *Applied Mathematics and Computation*, vol. 219, no. 20, pp. 10302–10312, 2013.

7 Combination of GIS and SHM in Prognosis and Diagnosis of Bridges in Earthquake-Prone Locations

Arman Malekloo
Middle East Technical University

Ekin Ozer
University of Strathclyde

Fadi Al-Turjman
Near East University

CONTENTS

7.1   INTRODUCTION

Earthquake as a natural disaster can effectively bring parts or all the transportation network systems, especially in metropolitan areas, to an immediate halt. Underestimating the seismic risks in bridges, one of the essential components of transportation infrastructures would bring chaos and disorder to the disaster areas. Bridges assist in transporting goods and disaster victims to and from cities and disaster sites. They are one of the elements in search and rescue in post-earthquake operations. Therefore, no proper analysis and assessment of the risk in bridges could undoubtedly cause disruptions to the transportation network and, ultimately, failure of the urban areas. This chapter investigates the use of structural health monitoring (SHM) and Geographical Information System (GIS) tools for mitigating the impacts of earthquake disasters on bridges at the response and recovery stages. What is more, it introduces a cloud-based framework which proposes the combined use of SHM-GIS as a tool to assess bridges and network systems in an improved and efficient manner compared to separate use of these items.

The efforts on the analysis of past events have considerably improved the resiliency of bridges to earthquakes, but there are still cases where they fail (Little, 2002). Moreover, bridges are considered spatially dispersed and interconnected structures. They are interdependent from each other; therefore, analyzing one bridge under seismic assessment would not necessarily provide enough information to propose suggestions and alternatives for the mitigation of future earthquakes. Moreover, although the current tools in the literature cover the basic requirements for bridge management mostly in a local manner, network-level assessment and decision-making platforms are still limited except few benchmark examples.

SHM is a monitoring technology that can detect damage and inspect the overall performance of structures, ideally in real time and in a continuous manner (Chang, 1998). Coupling SHM with forecast system performance, also known as damage prognosis (DP), can enable behavioral predictions to estimate the useful renaming time of the structures under future loads (Farrar and Lieven, 2007). Typically, SHM systems consist of arrays of sensors deployed on strategic locations on bridges that can collect critical spatial information such as vibrations and displacement. As discussed earlier, the need for assessing multiple bridges on the network is essential to produce effective countermeasures; however, collocated inclusion of multiple bridge monitoring systems and their effect on transportation network will result in a considerable amount of data that is hard to capture, analyze, and manage. This is where GIS comes into the picture.

GIS and its core functionality, i.e., organizing information in a standard graphical view (Tomaszewski, 2014), can aid SHM to represent better and manage the captured data. Therefore, a synergistic combination of the two systems can provide a decision-making platform to better decide on the suggestions and alternatives in disaster management and mitigation. GIS is the perfect platform when trying to analyze and show the impact of the failure of bridges on the transportation network in terms of functionality loss such as traffic delays and lack of connectivity, and economic loss in terms of local and regional levels.

The new paradigm shift in the Internet of Things (IoT) has led to many new innovative use cases of SHM. One of the examples is the utilization of drones for monitoring critical infrastructure. Drones or unmanned aerial vehicle (UAV) with its existing hardware, such as digital cameras, motion sensors, and communication units, can already contribute to SHM applications with minimal effort. Installing custom sensors such as vibration-based non-destructive testing (NDE) method can be useful for damage detection situations such as identifying damage in small and unreachable areas. Similar principles apply to mobile devices due to their multisensory environment and advanced computer skills (Alavi and Buttlar, 2019; Ozer, 2016). Such new innovative technology is considered as part of the shift towards cyber-physical SHM system (Ozer and Feng, 2019). Structures as the physical objects, cloud-based real-time engineering computation as the cyberobjects, and the sensors as the connecting medium present a modern infrastructure assessment and management scheme.

This chapter presents background and necessary information about SHM and GIS in Section 7.1. Sections 7.2 and 7.3 address standalone applications of each of the tools throughout the literature. Section 7.4 introduces the intelligent and sustainable cloud-based SHM-GIS framework for risk assessment of bridges in earthquake-prone locations. Machine learning (ML) as a complementary addition to SHM and GIS is explained in Section 7.5. Mobile device applications are still limited; therefore, drone-based SHM implications are discussed in Section 7.6. Finally, Section 7.7 concludes and highlights the future initiatives.

7.2 OVERVIEW OF THE TOOLS

This section discusses an overview of the SHM and GIS and their combined use in seismic performance assessment of the bridges. Although SHM systems mainly circle civil/mechanical/aerospace infrastructures, GIS has a vast list of applications and is an integral component of decision-making in many disciplines (Chrisman, 1999).

7.1.1 STRUCTURAL HEALTH MONITORING (SHM)

NDE for damage detection or identification through a series of sensors (either stationary or mobile) placed on a structure refers to as SHM. A vertical hierarchy is typically considered in order to identify damages. A pioneered damage typology scheme was offered by Rytter (1993). Damage state was categorized into four levels, namely,

1. Existence of damage – detection
2. Position of damage – location
3. Severity of damage – extent
4. Prognosis of damage – prediction.

In such a hierarchy, knowledge of the previous level is required for complete damage identification. This means that the success at each level depends on how well the lower levels perform. Damage could relate to any changes in the structural behavior of a structure that can change its current or future performance. By definition, change refers to a baseline that makes damaged and intact states comparative (Farrar and Worden, 2007). Many works have reviewed SHM applications in variety of disciplines such as (Arcadius Tokognon et al., 2017; Feng and Feng, 2018; Sony et al., 2019). The four-stage damage identification is the center of every SHM application. As shown in Figure 7.1, the SHM comprises many other elements and features.

In SHM paradigm, we first need to answer the following questions and carry out the procedures defined below (Farrar et al., 2001):

1. Why there is a need to evaluate damages and damage description? (Operational evaluation.)
2. Which quantities need to be selected and measured, which type of sensors are required, and how often the data should be collected? (Data acquisition.)
3. Extracting low-dimensional feature vectors and excluding redundant information in addition to data condensation. (Feature selection.)
4. Verifying the significance of the extracted feature using statistical analysis. (Statistical model development. (Feature discrimination.)

Conventional sensors used in SHM are accelerometers, strain gauges, corrosion sensors, fiber optic sensors (Noel et al., 2017), camera image/video processing (Yang et al., 2017), and many more. Deployment of these sensors requires one to determine the best optimal locations along the span or piers of a bridge since measuring, for instance, and vibration needs multiple placements of accelerometers to start the modal analysis of the bridge. Many algorithms could be employed in optimal sensor placement (OSP) techniques to identify these critical locations. Genetic algorithm for OSP of a long-span railway steel bridge in Deshan et al. (2011) and modified

FIGURE 7.1 SHM domain.

variance (MV) method described in Chang and Pakzad (2014) for Northampton Street and Golden Gate Bridge are among many of the examples in OSP studies. In addition to these, environmental factors such as weather conditions and fluctuations in temperature should also be taken into account as some sensors may have limitations under harsh conditions (Sohn et al., 2003). A wireless sensor network (WSN)-based SHM system architecture is shown in Figure 7.2.

7.1.2 Geographical Information System (GIS)

GIS constitutes many aspects, as shown in Figure 7.3. A visually explanatory platform involving GIS manages multiple data from different sources on separate layers allowing simulation and modeling of all data and their influence on one another. GIS and its useful applications in many disciplines, especially in disaster management cases, comprise shortcomings. The time, effort, and possibly money that are essential for advanced GIS may deter usage of the tool completely. Applicability constraints can clearly be seen when analyzing earthquake disasters and its implication on the network, and it can make the processing and analyzing, a complicated and time-consuming process (Tomaszewski, 2014).

GIS maps with different layers are available online[1]; however, the currency of the information provided may be of concern. Therefore, in some cases where there is a lack of information on the GIS maps (e.g., unknown bridge locations or highway network information), one needs to spend hours to acquire these data and import them into the correct location on the maps.

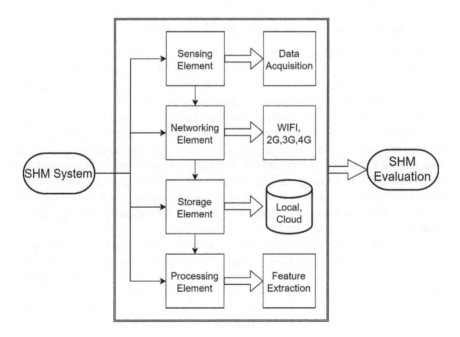

FIGURE 7.2 SHM system architecture.

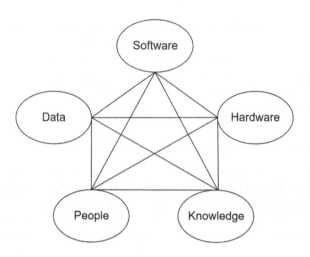

FIGURE 7.3   Components of GIS. (Adapted from Tomaszewski (2014, p. 75).)

7.3   BRIDGES PERFORMANCE ASSESSMENT

The subsequent sections review SHM and GIS applications for bridge management and monitoring. Both tools are discussed considering the features they possess solely based on their system architecture. In addition to standalone applications, a brief review of SHM-GIS applications is also presented.

7.3.1   Bridges Performance Assessment Using SHM

Accelerometers are widely used sensors in SHM systems due to their low cost and easy installation, as well as their easy integration with other methods such as GPS for better accuracy in inverse structural dynamics. Meng, Dodson, and Roberts (2007) introduced their GPS-triaxial accelerometers approach for the structural response of the Wilford Bridge in Nottingham. Another similar study on a pedestrian bridge was conducted by Moschas and Stiros (2011). In both studies, time synchronization between GPS and accelerometers and the problem of different sampling rates of the systems which require sophisticated filtering techniques are some of the matters that need consideration in future studies. Other than using individual sensors deployed over a bridge, with the advent of smartphones, one can use the said devices to acquire vibration data from citizens' smartphones in the paradigm of crowdsourcing applications (Ozer and Feng, 2016, 2017).

As with the development in technology and an increase in the complexity of human-made civil structures, there is a need for a more efficient and long-term solution for some of the conventional sensors used in today's SHM applications (Casas and Cruz, 2003). Optical fiber sensors (OFSs) provide improved quality of data acquisition, reliability, easy installation, and lower lifetime cost (Lopez-Higuera et al., 2011). The fidelity of OFSs in large and critical SHM systems often prevails over the high initial investment costs.

Another recent advancement in the SHM application is the use of digital video cameras with computer vision algorithms to identify displacement and vibration values. Specifically, in inaccessible locations on bridges, a contact-less vision approach is proven to be active and flexible in extracting information than other methods (Feng and Feng, 2017). In a study by Ye et al. (2013), charge-coupled device (CCD) digital camera with extended zoom up to 100m was performance-checked on Tsing Ma bridge, and the results were compared to MTS[2] 810 material testing system. In another research (Khuc and Catbas, 2017), displacement of a bridge was tested in a non-contact vision approach, and the difference in the results was less than 5% from using conventional sensors

7.3.2 BRIDGES PERFORMANCE ASSESSMENT USING GIS

GIS, as explained earlier, can be considered as a database management system capable of storing, analyzing, and displaying such data in a standard graphical interface. In the area of bridge performance assessment, standalone applications of GIS mostly concentrate on risk assessment and life-cycle risk analysis. Spatially distributed information along with multiple independent parameters of bridges and networks calls for a management system that could operate and analyze under different scenarios. Integrating bridge inventory information with earthquake parameters required to produce fragility curves to determine bridge damage state as the input parameter for initializing spatial analysis is widely used in many studies. In a seismic risk assessment (SRA)-based methodology for the Shelby County, Tennessee (Werner et al., 2000), 384 bridges in the network were assessed from multiple pre-generated earthquake models. Then, the traffic delay output was utilized to estimate the economic loss of the highway system. In the study done in St. Louis metropolitan area (Enke et al., 2008), ArcView was used as a spatial tool for mapping, locating, and setting up earthquake scenarios to evaluate the indirect economic loss, which was more significant than direct loss. Cheng, Wu, Chen, and Weng (2009) introduced a bridge repair/rehabilitation decision-making model where ArcGIS was used to identify alternative routes where detours were placed to reroute traffic. Later the economic loss model including the rehabilitation cost as well as additional costs through redirecting traffic was constructed. The decision-making model at the end offered either the lowest cost or the shortest duration of repair.

Another use of GIS is in the life-cycle assessment (LCA) of bridges. Deterioration of bridges over their lifetime and external attributes such as environment and traffic can influence the service life. In their research, Babanajad et al. (2018) introduced an LCA framework for the U.S. Bridge Inventory, rating the inventory as a whole. Their Long-Term Bridge Performance Portal (LTBP) incorporated GIS and Google Maps to query multiple information. An overview of the bridge management system using GIS is shown in Figure 7.4.

7.3.3 BRIDGES PERFORMANCE ASSESSMENT USING SHM-GIS

Combining the above-said tools into one network, it provides not only the capabilities of the instruments alone but also extra useful features that would not have

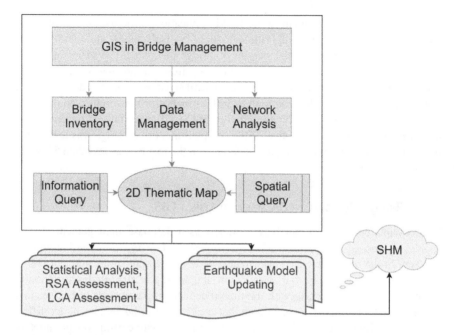

FIGURE 7.4   GIS system design for bridge management.

been attainable otherwise. In the study by (Jeong et al., 2017), a cloud-based cyber-infrastructure framework was presented. Apache Cassandra open-source as a column-oriented database and Microsoft Azure was used for the database management system and cloud provider, respectively. The web user interfaces for data extraction and information visualization on Google Maps were also provided. The Long-Term Bridge Performance (LTBP) Program[3], as also briefly mentioned in the previous section, is an initiative by the Highway Administration (FHWA) in 2008 envisioning a 20-year comprehensive field data collection from a sample of bridges in the U.S.A (Parvardeh et al., 2016). The program consists of analyzing bridge performance under deterioration. As previously discussed, the field data are gathered and maintained from different NDT techniques. The web-based platform containing the National Bridge Inventory (NVI) with GIS capability can enhance the quality of management of bridges by bridge owners as well as researchers to better understand the performance of bridges. A conceptualized GIS-based SHM was proposed in Shi et al. (2002). SQL Server database and Maptitude GIS were used to input and store bridge and sensor data and to visualize in an interface to view and extract the bridge/sensor information. Table 7.1 summarizes the above information and compares different features of bridge damage for risk assessment.

7.4   INTELLIGENT SHM-GIS CLOUD-BASED BRIDGE MONITORING SYSTEM

In this section, an intelligent SHM-GIS cloud-based framework is introduced and discussed. Individually deployed GIS and SHM tools discussed above may have their

TABLE 7.1
Comparison of Different Bridge Assessment Techniques

Application	Sensor(s) Requirement	Structural Damage Assessment	Database Management	Spatial Analysis	Decision-Making Feature	Risk Assessment Feature	Scalability	Cost
GIS	×	×	✓	✓	✓*	✓*	Low	Low
SHM	✓	✓	×	×	✓*	✓*	Medium	Medium
SHM-GIS	✓	✓	✓	✓	✓	✓	High	Medium

✓* indicates in limited scenarios given the use case of the application.

benefits in bridge monitoring, specifically for small-scale applications. However, in large-scale deployments considering the size and the complexity of the application, it may result in higher overall cost and less accurate results. It is, therefore, recommended that combined use of the tools on a cloud-based platform could enhance their performance. Furthermore, cloud-based platforms allow collaboration of different stakeholders enabling each party to add and edit on top of the existing data, and performing simultaneous analysis.

The nature of SHM systems in terms of sustainability already considers the three components of a sustainable approach, i.e., economic, social, and environmental. These include a reduction in traffic delays and downtimes, which subsequently lead to lower carbon dioxide emission and, lastly, the expected economic loss. These are mainly related to the network aspect, but the same can be said to the structure itself, i.e., bridge. Bridge monitoring can provide useful information in terms of the remaining helpful time and any maintenance that may be necessary for the future, which can help minimize the costs and maximums the life expectancy by early retrofitting or reconstructing.

Similarly, GIS can help bridge managers to have a better understanding of the structures and their behavior under different conditions. It can deliver a decision-making platform (Bǎneş et al., 2010) for the risk assessment of bridges and their cascading failures on the network, thus offering a complete management tool that could provide sets of strategies depending on the application use.

The new paradigm shift to cloud computing and web-based applications marks the SHM-GIS cloud-based platform a necessity in today's technological world. Not only it provides the core functionality of the tools, but instead, it goes further to expand its roots for even more cost-effective, efficient, and sustainable solution in bridge prognosis and diagnosis. Synergistic use of SHM and GIS can develop or update earthquake models on the fly and provide a more accurate damage estimate of the bridge and its effect on the network.

The proposed SHM-GIS cloud-based system architecture is, therefore, presented in Figure 7.5. The sensory subsystem layer acts as the data acquisition where it collects the data from bridges. The collected information is then transferred to a server via a different form of communication standards such as Wi-Fi, Bluetooth, and cellular and later uploaded on the cloud. Due to the enormous size of the acquired data for any given time history chiefly in the extensive application of bridge monitoring, storage methods need investigation. The issue of big data and storage has led to the creation of different file structure format. Standard file formats for storing large amounts of data are (1) HDF[4] (Hierarchical Data Format) and (2) netCDF[5] (Network Common Data Format). However, due to the file structure of these formats, they are not ideal in cloud computing. Many alternatives with their strengths and weaknesses are present in Matthew Rocklin (2018) webpage. HDF5 (Hierarchical Data Format version 5) can be an ideal solution in this case for storing multi-dimensional data. Bridge information such as geometry and location, and network description such as highway information and traffic information are stored in an object-related database management system (ORDBMS). PostgreSQL, with the extension, *PostGIS* for handling spatial data, is the common database management system (DBMS) for

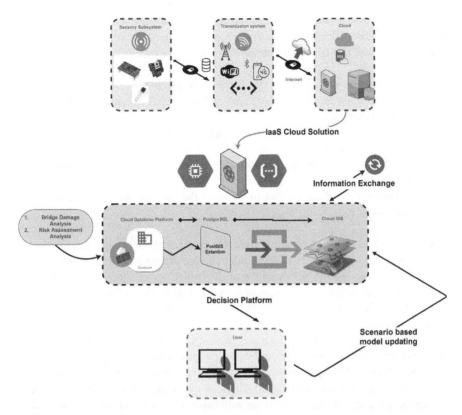

FIGURE 7.5 Framework of the system architecture.

SHM applications. PostgreSQL is an open-source DBMS that is well developed and intuitive. The relationship between the sensor data and the structural/network elements is also a one-to-many relation.

The cloud service for this system relies on infrastructure as a service (IaaS) type. IaaSs are often low-cost, more accessible, and faster options over different cloud services enabling storage resiliency, frequent backup, and high level of automation. Deciding which cloud provider to use depends on the performance and uptime required from the provider. A typical solution for cloud computing is Google Cloud and Compute Engine. Other services, such as Microsoft Azure and Amazon Web Services (AWS), are also available. These data then proceed into performance analysis and monitoring of the bridges. Depending on the data type (vibration, displacement, image, etc.), different algorithms can define the damage state in the given earthquake scenario.

Incorporating the network data such as traffic delay and routing info into the database can enable the employment of a cloud GIS platform capable of visualizing, analyzing, managing, and monitoring bridges and the effects of failure of them on the transportation network. Using this information and a simple risk formula that includes direct costs such as structural loss, network loss, and indirect loss, it can provide a decision-making platform for pre- and post-earthquake disaster scenarios. The advantages of this SHM-GIS cloud-based system are as follows:

- Utilizing open-source and free software and system providers
- Ability to add/remove or change any information without the problem of proof checking for errors
- The flexibility of the system in any application use (using a small or large number of sensors)
- An intuitive and low-cost solution for bridge monitoring (especially for bridges owners)
- The scalability of the system in terms of the location and the size of the application.

Moreover, risk assessment based on dynamic changes in the model can also serve in the system. As parameters of the model change throughout time, real-time risk assessment can assess the performance of the bridge under future loads. The data from traffic and future loading can predict the future state of the bridge, aiding bridge owners to decide about retrofitting or reconstructing all or some parts of bridge elements.

The whole system, from the data acquisition, DBMS, and user interface, can be programmed with the open-source Python programming language. Web applications, as well as mobile applications for viewing and extracting information, can also be implemented for easier and faster utilization of the data. The ability of information exchange and information sharing with other software and services is another advantage that distinguishes this from other similar systems (Ellenberg et al., 2015; Eschmann et al., 2012; Sankarasrinivasan et al., 2015). A summary of the traditional SHM-GIS damage assessment and the cloud-based variant is tabulated in Table 7.2. The next section brings a recent technological implementation and aerial devices, which provide an efficient synthesis of GIS and SHM domains.

7.5 MACHINE LEARNING IN SHM APPLICATION: A COMPLEMENTARY ADDITION

Given the amount of data gathered from many different things, it is important to understand the pattern that underlines it. Day by day with increase in complexity of structures, without automatic (sometimes semiautomatic) processes to discover patterns using computer, such tasks would be infeasible and impractical. ML is considered as tool to recognize/classify information based on a learned pattern through the use of different algorithms. In general, ML algorithms are based on either (1) statistical, (2) neural, or (3) synthetic approaches. The first two approaches are generally considered as the main pattern classifiers for SHM (Bǎneş et al., 2010). There are many works utilizing ML. For example, Cao et al. (2018) developed a piezoelectric impedance measurement for an effective structural damage identification through an inverse analysis. Similarly, in a study by Moore et al. (2012), crack identification in a thin plate was achieved by model updating.

TABLE 7.2
Summary of the Traditional and New Novel SHM System

References	System	Real-Time Processing	Flexibility	System Efficiency	Mobility (Easy Accessibility)	Maintenance and Management	Open and Interoperable	Multi-Purpose Decision-Making
Jeong et al. (2017) and Shi et al. (2002)	Conventional SHM-GIS	X	Low	Medium	Low	Low	X	X
This study framework	Cloud-Based SHM-GIS	✓	High	High	High	Medium	✓	✓

With the advent of ML and statistical pattern recognition algorithms, a new level can be added to the Rytter (1993) four-stage damage identification. Type of damage or classification of damage is the level that is possible through the use of ML algorithms. This new step lies between step 2 and step 3 introduced by Rytter. To illustrate this, Figure 7.6 depicts the five-stage damage identification in SHM application given the domain and level of difficulty.

Given that both damage and undamaged information are available, a supervised learning algorithm can effectively go through all five levels of damage detection. This requires many data to be readily available from the sensing systems or the physical-based models and the experiments. This is not possible in many applications, and the current information for damage sate is limited, if not, unavailable. For such situations, there exists a method called unsupervised learning. In this mode, instead of learning the models and train based on the data, a rather simple approach, novelty, or outlier detection is applied (Casas and Cruz, 2003). Figure 7.7 illustrates a statistical pattern recognition model for a typical damage assessment scenario utilizing ML. Moreover, Table 7.3 shows the current reviews on ML utilization on SHM application.

ML can augment SHM in many aspects which the old system is incapable of. For example, environmental and operational variabilities oftentimes are not considered but have proven that they can greatly influence in-service structures (Sohn, 2007). Including these effects by leveraging the power of ML can definitely help SHM application achieve better level of detection. Moreover, ML and deep learning can be particularly useful in bridge monitoring applications which are combined with GIS and remote sensing tools that utilize machine vision for anomaly detection or as tools in data analytics inside the GIS package.

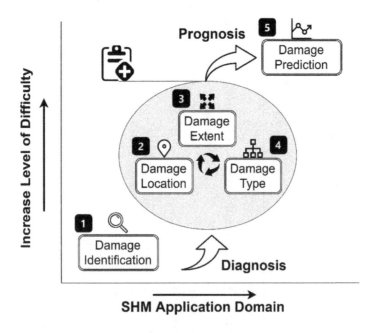

FIGURE 7.6 Five-stage damage identification.

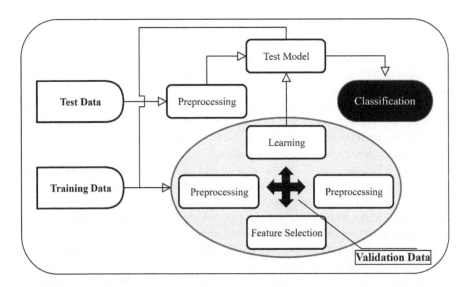

FIGURE 7.7 A typical ML model.

TABLE 7.3
Works on ML Utilization in SHM

Reference	Model-Based	Data-Based	Application of ML/Deep Learning	Mobile Applications	Machine Vision Consideration	Novel Applications (UAV, VR, AR, etc.)
Fan and Qiao, (2010) and Gomes et al. (2019)	✓	-	✓	-	-	-
Ye et al. (2016)	-	✓	-	-	✓	-
An et al. (2013)	✓*	✓*	✓*	-	-	-
Moughty and Casas (2017)	✓*	✓	✓*	-	-	-
Kerle et al. (2019)	-	-	✓	-	-	UAV Only

✓* indicates a little information.

7.6 DRONE-ASSISTED SHM: A SYNERGISTIC MEDIUM

Drone technology, also known as unmanned aerial vehicle (UAV), has seen a vast increase in usage in recent years due to the advantages they can offer and especially their deployment flexibility (Al-Turjman et al., 2019). Given their versatility, low-cost as well as ease of deployment elements of a flying piece of technology, they are becoming more and more accretive (Al-Turjman et al., 2020). There are a limited number of studies focusing on drone-based SHM systems. Most of the works aim at the post-image processing of cracks (Ellenberg et al., 2015; Eschmann et al., 2012; Sankarasrinivasan et al., 2015).

However, very few have focused on vibration-based SHM (Na and Baek, 2016), but with the advancement in ML and cloud computing, image processing on the fly augmented with innovative technologies is considered the next step in mobile SHM applications. Some works have already started extracting critical information from the drones (Hoskere et al., 2019), but majority of them rely on post-processing techniques. Using the framework introduced, with the total flexibility it offers, application of cloud computing can become a reality. With the GIS part of the framework, critical data points on the structure can be generated and regularly visited to detect any abnormal changes with respect to a baseline.

Moreover, with the combined use of SHM-GIS in drone-based SHM applications, both on-fly and cloud processing of information can be achieved, and immediate results can be shown as a map. The other benefit that this system also offers is the well-regulated and controlled behavior. This in turn provides total control over how the system should be implemented for the most effective use of drones.

7.7 CONCLUSION

Bridges are indispensable to a transportation network. Earthquakes can damage bridges and effectively disrupt the transportation network. SHM and GIS are some of the tools that can mitigate, understand, and manage these issues. In this chapter, a cloud-based SHM-GIS framework targets bridge monitoring in earthquake-prone locations.

SHM and GIS both have their advantages in the application of monitoring and management of bridges. However, by synergistically combining these tools, researchers and especially owners of bridges can utilize the mixed results to have a better understanding of bridge performance with a decision-making platform extension. The new paradigm shift to cloud computing has enabled us to offload both database and data computations to a cloud server. Cloud computing can increase productivity, speed, and security of data by doing so in a low-cost manner. By enabling this feature, we can utilize the power of the cloud to visualize, analyze, manage, and store multiple data from multiple bridges.

Application of cloud GIS can help to envision what would happen under different earthquake scenarios and what would be direct and indirect losses of such an event. Besides, with the help of this system, damage prediction for future events such as an increase in traffic load or deterioration of bridges and the implications of it on the transportation network can be examined.

The proposed framework uses free and open-source software and packages, and can introduce a web or mobile-based application written in Python alone. With the help of the information exchange feature of the system, the beneficiaries can extract data and use them in other services or software with little to no modification. Also, this chapter introduced ML as a complementary addition and the use of drones as a synergistic medium to SHM application that can be included in the proposed framework for the most effective implementation of prognosis and diagnosis of bridges and bridge monitoring in general.

The concept provided in this chapter, with its flexible and open-source items, can be considered as the next step towards the future of the cyber-physical system (CPS) with many new features as part of the IoT paradigm shift (Al-Turjman and Malekloo, 2019). The future of the risk assessment for transportation network lies within the cloud. What will ensue from such a movement towards this paradigm are the applications of deep learning, artificial intelligence, drones, and virtual/augmented reality. These are a tiny droplet in the vast ocean of the next generation sensing and monitoring applications.

As future work, the focus will be on the implementation and development of such a system in addition to including dynamic and real-time risk assessment procedures embedded into the system for further performance analysis (Malik et al., 2020).

NOTES

1 Natural Earth Data, Esri Open Data, USGS Earth Explorer, OpenStreetMap.
2 http://www.mts.com/en/index.htm
3 https://fhwaapps.fhwa.dot.gov/ltbpp/
4 http://www.hdfgroup.org/
5 https://www.unidata.ucar.edu/software/netcdf/

REFERENCES

Alavi, A.H., Buttlar, W.G., 2019. An overview of smartphone technology for citizen-centered, real-time and scalable civil infrastructure monitoring. *Future Gener. Comput. Syst.* 93, 651–672. https://doi.org/10.1016/j.future.2018.10.059.

Al-Turjman, F., Abujubbeh, M., Malekloo, A., 2019. Deployment Strategies for Drones in the IoT Era: A Survey, in: *Drones in IoT-Enabled Spaces*. Boca Raton, FL : CRC Press/ Taylor & Francis Group, 2019, pp. 7–42. https://doi.org/10.1201/9780429294327-2.

Al-Turjman, F., Abujubbeh, M., Malekloo, A., Mostarda, L., 2020. UAVs assessment in software-defined IoT networks: An overview. *Comput. Commun.* 150, 519–536. https://doi.org/10.1016/j.comcom.2019.12.004.

Al-Turjman, F., Malekloo, A., 2019. Smart parking in IoT-enabled cities: A survey. *Sustain. Cities Soc.* 101608. https://doi.org/10.1016/j.scs.2019.101608

An, D., Choi, J.H., Kim, N.H., 2013. Options for Prognostics Methods: A review of data-driven and physics-based prognostics, in: *54th AIAA/ASME/ASCE/AHS/ASC Structures, Structural Dynamics, and Materials Conference*. p. 1940.

Arcadius Tokognon, C., Gao, B., Tian, G.Y., Yan, Y., 2017. Structural health monitoring framework based on internet of things: A survey. *IEEE Internet Things J.* 4, 619–635. https://doi.org/10.1109/JIOT.2017.2664072.

Babanajad, S., Bai, Y., Wenzel, H., Wenzel, M., Parvardeh, H., Rezvani, A., Zobel, R., Moon, F., Maher, A., 2018. Life cycle assessment framework for the U.S. bridge inventory. *Transp. Res. Rec. J. Transp. Res.* Board 036119811878139. https://doi.org/10.1177/0361198118781396

Băneş, A., Orboi, M.D., Monea, A., Monea, M., 2010. Sustainable development by GIS. *Res. J. Agric. Sci.* 42, 405–407.

Cao, P., Qi, S., Tang, J., 2018. Structural damage identification using piezoelectric impedance measurement with sparse inverse analysis. *Smart Mater. Struct.* 27, 035020. https://doi. org/10.1088/1361-665X/aaacba.

Casas, J.R., Cruz, P.J.S., 2003. Fiber optic sensors for bridge monitoring. *J. Bridge Eng.* 8, 362–373. https://doi.org/10.1061/(ASCE)1084-0702(2003)8:6(362).

Chang, F.-K., 1998. Structural Health Monitoring: A Summary Report on the First Stanford Workshop on Structural Health Monitoring, September 18–20, 1997 (No. SPO-18042). STANFORD UNIV CA.

Chang, M., Pakzad, S.N., 2014. Optimal sensor placement for modal identification of bridge systems considering number of sensing nodes. *J. Bridge Eng.* 19, 04014019. https://doi. org/10.1061/(ASCE)BE.1943-5592.0000594.

Cheng, M.-Y., Wu, Y.-W., Chen, S.-J., Weng, M.-C., 2009. Economic evaluation model for post-earthquake bridge repair/rehabilitation: Taiwan case studies. *Autom. Constr.* 18, 204–218. https://doi.org/10.1016/j.autcon.2008.08.004.

Chrisman, N.R., 1999. What Does 'GIS' Mean? *Trans. GIS* 3, 175–186. https://doi. org/10.1111/1467-9671.00014.

Deshan S., Zhenhua W., Li Q., 2011. Optimal Sensor Placement for Long-span Railway Steel Truss Cable-stayed Bridge, in: 2011 *Third International Conference on Measuring Technology and Mechatronics Automation.* Presented at the 2011 International Conference on Measuring Technology and Mechatronics Automation (ICMTMA), IEEE, Shangshai, pp. 795–798. https://doi.org/10.1109/ICMTMA.2011.482

Ellenberg, A., Branco, L., Krick, A., Bartoli, I., Kontsos, A., 2015. Use of unmanned aerial vehicle for quantitative infrastructure evaluation. *J. Infrastruct. Syst.* 21, 04014054. https://doi.org/10.1061/(ASCE)IS.1943-555X.0000246.

Enke, D.L., Tirasirichai, C., Luna, R., 2008. Estimation of earthquake loss due to bridge damage in the St. Louis Metropolitan Area. II: Indirect losses. *Nat. Hazards Rev.* 9, 12–19. https://doi.org/10.1061/(ASCE)1527-6988(2008)9:1(12).

Eschmann, C., Kuo, C.-M., Kuo, C.-H., Boller, C., 2012. Unmanned aircraft systems for remote building inspection and monitoring, in: *Proceedings of the 6th European Workshop on Structural Health Monitoring, Dresden*, Germany, pp. 1179–1186.

Fan, W., Qiao, P., 2010. Vibration-based damage identification methods: A review and comparative study. *Struct. Health Monit.* 83–11. https://doi.org/10.1177/1475921710365419

Farrar, C.R., Doebling, S.W., Nix, D.A., 2001. Vibration–based structural damage identification. *Philos. Trans. R. Soc. Lond. Ser. Math. Phys. Eng. Sci.* 359, 131–149. https://doi. org/10.1098/rsta.2000.0717.

Farrar, C.R., Lieven, N.A.J., 2007. Damage prognosis: The future of structural health monitoring. *Philos. Trans. R. Soc. Math. Phys. Eng. Sci.* 365, 623–632. https://doi.org/10.1098/ rsta.2006.1927.

Farrar, C.R., Worden, K., 2007. An introduction to structural health monitoring. *Philos. Trans. R. Soc. Math. Phys. Eng. Sci.* 365, 303–315. https://doi.org/10.1098/rsta.2006.1928.

Feng, D., Feng, M.Q., 2017. Experimental validation of cost-effective vision-based structural health monitoring. *Mech. Syst. Signal Process.* 88, 199–211. https://doi.org/10.1016/j. ymssp.2016.11.021.

Feng, D., Feng, M.Q., 2018. Computer vision for SHM of civil infrastructure: From dynamic response measurement to damage detection – A review. *Eng. Struct.* 156, 105–117. https://doi.org/10.1016/j.engstruct.2017.11.018.

Gomes, G.F., Mendez, Y.A.D., da Silva Lopes Alexandrino, P., da Cunha, S.S., Ancelotti, A.C., 2019. A review of vibration based inverse methods for damage detection and identification in mechanical structures using optimization algorithms and ANN. *Arch. Comput. Methods Eng.* 26, 883–897. https://doi.org/10.1007/s11831-018-9273-4.

Hoskere, V., Park, J.-W., Yoon, H., Spencer, B.F., 2019. Vision-based modal survey of civil infrastructure using unmanned aerial vehicles. *J. Struct. Eng.* 145, 04019062. https://doi.org/10.1061/(ASCE)ST.1943-541X.0002321.

Jeong, S., Hou, R., Lynch, J.P., Sohn, H., Law, K.H., 2017. A distributed cloud-based cyber-infrastructure framework for integrated bridge monitoring, in: Lynch, J.P. (Ed.), *Presented at the SPIE Smart Structures and Materials + Nondestructive Evaluation and Health Monitoring*, Portland, Oregon, United States, p. 101682W. https://doi.org/10.1117/12.2270716.

Kerle, N., Nex, F., Gerke, M., Duarte, D., Vetrivel, A., 2019. UAV-based structural damage mapping: A review. *ISPRS Int. J. Geo-Inf.* 9, 14. https://doi.org/10.3390/ijgi9010014.

Khuc, T., Catbas, F.N., 2017. Computer vision-based displacement and vibration monitoring without using physical target on structures. *Struct. Infrastruct. Eng.* 13, 505–516. https://doi.org/10.1080/15732479.2016.1164729.

Little, R.G., 2002. Controlling cascading failure: Understanding the vulnerabilities of interconnected infrastructures. *J. Urban Technol.* 9, 109–123. https://doi.org/10.1080/106307302317379855.

Lopez-Higuera, J.M., Cobo, L.R., Incera, A.Q., Cobo, A., 2011. Fiber optic sensors in structural health monitoring. *J. Light. Technol.* 29, 587–608. https://doi.org/10.1109/JLT.2011.2106479.

Malik, K., Ahmad, M., Khalid, S., Ahmad, H., Al-Turjman, F., Jabbar, S., 2020. Image and Command Hybrid Model for Vehicles Control using Internet of Vehicles (IoV). *Wiley Trans. Emerg. Telecommun. Technol.*, vol. 31, no. 5 .

Meng, X., Dodson, A.H., Roberts, G.W., 2007. Detecting bridge dynamics with GPS and triaxial accelerometers. *Eng. Struct.* 29, 3178–3184. https://doi.org/10.1016/j.engstruct.2007.03.012

Moore, E.Z., Nichols, J.M., Murphy, K.D., 2012. Model-based SHM: Demonstration of identification of a crack in a thin plate using free vibration data. *Mech. Syst. Signal Process.* 29, 284–295. https://doi.org/10.1016/j.ymssp.2011.09.022

Moschas, F., Stiros, S., 2011. Measurement of the dynamic displacements and of the modal frequencies of a short-span pedestrian bridge using GPS and an accelerometer. *Eng. Struct.* 33, 10–17. https://doi.org/10.1016/j.engstruct.2010.09.013

Moughty, J.J., Casas, J.R., 2017. A state of the art review of modal-based damage detection in bridges: Development, challenges, and solutions. *Appl. Sci.* 7, 510. https://doi.org/10.3390/app7050510.

Na, W., Baek, J., 2016. Impedance-based non-destructive testing method combined with unmanned aerial vehicle for structural health monitoring of civil infrastructures. *Appl. Sci.* 7, 15. https://doi.org/10.3390/app7010015.

Noel, A.B., Abdaoui, A., Elfouly, T., Ahmed, M.H., Badawy, A., Shehata, M.S., 2017. Structural health monitoring using wireless sensor networks: A comprehensive survey. *IEEE Commun. Surv. Tutor.* 19, 1403–1423. https://doi.org/10.1109/COMST.2017.2691551.

Ozer, E., 2016. *Multisensory Smartphone Applications in Vibration-Based Structural Health Monitoring* (PhD Thesis). New York, NY, USA: Columbia University.

Ozer, E., Feng, M.Q., 2016. Synthesizing spatiotemporally sparse smartphone sensor data for bridge modal identification. *Smart Mater. Struct.* 25, 085007. https://doi.org/10.1088/0964-1726/25/8/085007.

Ozer, E., Feng, M.Q., 2017. Direction-sensitive smart monitoring of structures using heterogeneous smartphone sensor data and coordinate system transformation. *Smart Mater. Struct.* 26, 045026. https://doi.org/10.1088/1361-665X/aa6298.

Ozer, E., Feng, M.Q., 2019. Structural reliability estimation with participatory sensing and mobile cyber-physical structural health monitoring systems. *Appl. Sci.* 9, 2840. https://doi.org/10.3390/app9142840.

Parvardeh, H., Babanajad, S., Ghasemi, H., Maher, A., Gucunski, N., Zobel, R., 2016. The Long-Term Bridge Performance (LTBP) Program Bridge Portal, in: *Proceedings of the International Conference on Road and Rail Infrastructure CETRA. Presented at the 4th International Conference on Road and Rail Infrastructure – CETRA 2016.*

PostGIS — Spatial and Geographic Objects for PostgreSQL [WWW Document], n.d. URL https://postgis.net/ (accessed 12.16.18).

PostgreSQL: The world's most advanced open source database [WWW Document], n.d. URL https://www.postgresql.org/ (accessed 12.16.18).

Rocklin, M., 2018. HDF in the Cloud [WWW Document]. URL https://matthewrocklin.com/blog//work/2018/02/06/hdf-in-the-cloud (accessed 12.16.18).

Rytter, A., 1993. *Vibrational based inspection of civil engineering structures* (PhD Thesis). Dept. of Building Technology and Structural Engineering, Aalborg University.

Sankarasrinivasan, S., Balasubramanian, E., Karthik, K., Chandrasekar, U., Gupta, R., 2015. Health monitoring of civil structures with integrated UAV and image processing system. *Procedia Comput. Sci.* 54, 508–515. https://doi.org/10.1016/j.procs.2015.06.058

Shi, W., Cheng, P.G., Ko, J.M., Liu, C., 2002. GIS-based bridge structural health monitoring and management system, in: Gyekenyesi, A.L., Shepard, S.M., Huston, D.R., Aktan, A.E., Shull, P.J. (Eds.), *Presented at the NDE For Health Monitoring and Diagnostics,* San Diego, CA, pp. 12–19. https://doi.org/10.1117/12.470725.

Sohn, H., 2007. Effects of environmental and operational variability on structural health monitoring. *Philos. Trans. R. Soc. Math. Phys. Eng. Sci.* 365, 539–560. https://doi.org/10.1098/rsta.2006.1935.

Sohn, H., Farrar, C.R., Hemez, F.M., Shunk, D.D., Stinemates, D.W., Nadler, B.R., Czarnecki, J.J., 2003. *A Review of Structural Health Monitoring Literature: 1996–2001.* Los Alamos National Laboratory.

Sony, S., Laventure, S., Sadhu, A., 2019. A literature review of next-generation smart sensing technology in structural health monitoring. *Struct. Control Health Monit.* 26, e2321. https://doi.org/10.1002/stc.2321

Tomaszewski, B., 2014. *Geographic Information Systems (GIS) for Disaster Management,* 1 ed. Boca Raton, FL: Routledge.

Yang, Y., Dorn, C., Mancini, T., Talken, Z., Kenyon, G., Farrar, C., Mascareñas, D., 2017. Blind identification of full-field vibration modes from video measurements with phase-based video motion magnification. *Mech. Syst. Signal Process.* 85, 567–590. https://doi.org/10.1016/j.ymssp.2016.08.041.

Ye, X.W., Dong, C.Z., Liu, T., 2016. A review of machine vision-based structural health monitoring: Methodologies and applications. *J. Sens.* 2016, 1–10. https://doi.org/10.1155/2016/7103039.

Ye, X.W., Ni, Y.Q., Wai, T.T., Wong, K.Y., Zhang, X.M., Xu, F., 2013. A vision-based system for dynamic displacement measurement of long-span bridges: Algorithm and verification. *Smart Struct. Syst.* 12, 363–379. https://doi.org/10.12989/sss.2013.12.3_4.363.

8 Smart Medium Access in Mobile IoT

Fadi Al-Turjman
Near East University

CONTENTS

8.1 INTRODUCTION

Smart cites are becoming not only a reality but also more popular and spreading day by day. In fact, the deployment of smart and cognitive cities is expected to be very dominant in several regions of the world in the near future due to their advantages and merits in making our life easier, faster, simpler, and safer [1–4]. However, still thousands of people die in car accidents/collisions every year. For example, over 1.2 million were died, for example, in traffic accidents around the world in 2016 [1]. There are a number of different reasons of road accidents/collisions in different countries. Nevertheless, the major cause of these road collisions is mainly taking the car under unpredicted weather conditions. In Ref. [1], the authors note that fifty percent of the fatal collisions happens while driving under a speed of 55 km/h. We therefore need a system where the speed limits are set according to the current weather, traffic, and road conditions.

Smart cities are composed of massive amount of smart, intelligent devices that have sensing, computing, actuating, and communication capabilities. These devices are designed in such a way that reduces human intervention through cognition and automation in communication. They are commonly termed as Internet of Things (IoT) devices, which are expected to dominate the infrastructure of smart cities. Among the massive design requirements of IoT devices, low complexity with highly reliable communication channels comes as a key priority besides energy efficiency, latency, and security in emerging 5G services [5]. Of course, the impact of such an enabling technology is expected to be revolutionary. The new infrastructure for communication

159

is expected to transform the world of connected sensors and reshape several existing industries. Such a revolution would of course require extensive research and development for the co-existence and device inter-operability with 5G systems, and deployed sensor networks. This is a must for cooperative and smart sensing techniques, improved quality of service (QoS), and energy-efficient architectures in new intelligent transportation system (ITS) infrastructure. This can be achieved by developing new sensor/5G protocols and standards, which can communicate with the cloud reliably.

Focusing on optimization of the ways the data is exchanged between the sensory devices and applications in IoT and cyber-physical systems has led to the 5G/IoT integration as well. For example, studies on approaches to construct higher-level abstractions of data at local gateways are proposed to reduce the traffic load imposed on the communication networks that provide the real-world data [5]. Test beds and real-world data sets are popularly used to analyze the methods proposed.

With this recent revolution in wireless telecommunications, several advanced solutions relying on wireless communication standers have been proposed to provide intelligent transportation systems in the IoT paradigm. For instance, in Ref. [2], the authors projected an automatic speed control (ASC) system that adjusts the speed of the vehicle according to the speed limit on the road. The feasibility of a smart box called "telematics," which has the ability to capture, analyze, and communicate, is being studied in cooperation with IBM's Engineering and Technology Services. Using multiple microprocessors and tiny sensors attached to the vehicle body, it is able to observe the vehicle's velocity, for example, and compare it to the upper speed limit of the road. In case the speed of the car is higher than the announced limit allowed by authorities, the box will verbally notify the driver. Moreover, a digital image processing system has been proposed by Baró et al. [3] while utilizing onboard cameras to read and recognize signs at the side of the road and send the warning signal to the driver and/or directly control the car. Different versions of this system have been investigated intensively all over the world.

Results in Ref. [4] have shown that this solution is able to cut down accidents rate by 35%. In the near future, the speed control system will be very dependent on the standard of IEEE 802.16 to locate each vehicle and satisfy the demands of the required real-time services such as voice and video [5].

In IEEE 802.16, there are different medium access control (MAC) scheduling services, such as UGS (Unsolicited Grant Service), rtPS (real-time polling service), and nrtPS (non-real-time polling service) to provide better quality of services [26][27]. There are two commonly used schedulers for real-time traffic: the UGS and the rtPS [6]. However, these schedulers do not fulfill the requirements for the real-time services in smart cities. Hence, the suitability of a batch Markovian arrival process (BMAP) is analyzed [7] for modeling of IP-based data traffic. Accordingly, the BMAP has been found to be better in comparison with Markovian-based models. Hence, we propose a real-time-BMAP (RT-BMAP) model for real-time services. The objective of this approach is to provide the required QoS level with the minimum delay. Accordingly, the major contributions of this chapter are as follows:

1. We present the enhanced real-time polling system (E-rtPS) integrated with RT-BMAP to solve the interference problem in smart cities paradigm optimistically.

2. Our proposed approach has less computational overhead and better performance in terms of resource utilization.
3. The examination of real-time results shows that the proposed framework outperforms the existing IEEE 802.16 services in terms of delay, throughput, and reception percentages.

For more readability, used abbreviations along with their definitions are provided in Table 8.1. The rest of this chapter is organized as follows. Section 8.2 overviews related works in the literature. A detailed description of the proposed framework is

TABLE 8.1
Used Abbreviations and Acronyms

Abbreviations	Definitions
IoT	Internet of Things
BMAP	Batch Markovian arrival process
RT-BMAP	Real-time batch Markovian arrival process
MAC	Medium access control
UGS	Unsolicited grant service
nrtPS	Non-real-time polling service
rtPS	Real-time polling service
QoS	Quality of service
ASC	Automatic speed control
ITS	Intelligent transportation systems
LTE-A	Long-Term Evolution-Advanced
E-rtPS	Enhanced real-time polling service
SNR	Signal-to-noise ratio
WSN	Wireless sensor network
BS	Base station
RF	Radio frequency
OFDM	Orthogonal frequency division medium
MCC	Management and Control Center
PSTN	Public switched telephone network
IRSN	IP real-time subsystem network
GS	Grant size
MC	Mobile controllers
LC	Local controllers
VoIP	Voice over IP
MST	Minimum spanning tree
RS	Real-time server
RC	Real-time client
RSU	Road-Side Unit
BER	Bit error rate
FER	Frame Error Rate
CSCF	Call Session Control Function
HSS	Home Subscriber Station
D2D	Device-to-device communication

presented in Section 8.3. Extensive simulation results are discussed in Section 8.4. Finally, Section 8.5 is concluding the work in this chapter.

8.2 RELATED WORK

By 2021, the mobile networks will triple the amount of used smart devices in 2016 [28, 29]. In addition, mobile video services will increase nine times what we experienced in 2017. Given that our resources are limited and QoS is critical, the system base station (BS) shall allocate the existing bandwidth proactively. In smart cities, smart transportation systems, cellular network, wireless sensor network (WSN), device-to-device (D2D), and many of these smart paradigms involve the wireless mobile multimedia communications and/or transmissions.

Specifically, the D2D has received increased attention from wireless cellular networks because it can noticeably offload the network traffic and reduce the transmission energy since the communicating devices are in close proximity of each other [2]. This close proximity makes it an eminent networking architecture for relay-based transmission used by energy-constrained devices, such as smartphones, tablets, and PDAs. Therefore, it is considered to be vital to improve the energy efficiency of D2D networks while maintaining their reliability in terms of the experienced throughput and latency. Most of the existing works have considered the direct link situation for the degradation of following reasons, namely, long-distance communication [2], poor propagation [3], and interference [4].

As relay-based transmission can improve the longer distance D2D communication, most of the researchers have given their attention for the multi-hopping D2D communication. In Ref. [5], the authors deal with spatial density and power transmission to get better transmission capacity. However, it is declared without specifying a routing technique. In Ref. [8], the authors propose a multi-hop routing technique to maximize the hop count of the networking systems. But then, it is proven undependable to minimize the distance between the users. Since the existing relay-based techniques do not consider the energy efficiency for the D2D communication, there is a need to formulate a combinatorial optimization that improves the energy efficiency of such communication systems. Accordingly, technological advancements are balancing the act between bandwidth usage and time delay in any data delivery approach. In addition, low complexity with high reliability communication comes as a first key priority besides energy efficiency, latency, and security [5].

There are two main categorizes of approaches that are focusing on addressing these challenges: distributed approaches and centralized approaches. The distributed approach is a distributed routing in the network which is very popular, and it has many sub-approaches. Distributed coding is also proposed and adopted to be used in future 5G and beyond systems as indispensable channel coding scheme [9,10]. However, different channel codes are constructed and generated according to a pre-specified value of the signal-to-noise ratio (SNR). This issue becomes even more complicated with channel fading consideration due to multipath existence. In an attempt towards proposing reliable channel codes which considers fading conditions, several studies have been conducted and performed in the literature. In Ref. [11], a multi-level, fading channel custom that utilizes nested coding is exhibited. It was demonstrated that this approach tends to tackle fading channels with limited number of states. In Ref. [12], polar codes are connected to remote channels. A strategy for acquiring the Bhattacharyya parameters related to Rayleigh channels is exhibited. In Ref. [13], lower and upper limits on Bhattacharyya parameters were selected to develop competitive channel codes. In Ref. [14], the authors detailed the coding procedures for the radio frequency (RF) channel with realized channel state data at the two ends of the connection with known channel dissemination data. Another example on investigating coding for blockfading channels can be found in Ref. [15].

The channel use is exhibited with the objective that codes could be grasped to be encoded over isolated squares in Ref. [16]. In Ref. [17], creators analyze the general sort of the superfluous information trade and coding with multistage understanding. In light of this examination, they proposed a graphical structure methodology to create channel codes for picture hindrance. In Ref. [18], a clear system for the improvement of portable channel codes considering Rayleigh impacts was displayed. The sub-channels are shown as multipath obscuring channels, and their varying decent variety request and clamor contrast are pursued. In Ref. [19], codes are planned solely for square-based channels. The authors built channel codes customized for square fading channels while polarizing it with codes. The accomplished arrangements are shown to convey significant increase contrasted with ordinary coding.

In Ref. [20], the authors gave an elective structure of this methodology by treating the shadowing, fading, and coding of the channel as a solitary element. This empowers developing channel codes by mapping the code with shadowing and fading effects. The acquired codes adjust to divert vacillation in versatile situations. In Ref. [21], the author proposed a scheduling scheme of dynamic carrier aggregation (DCA) to provide higher energy efficiency in uplink communication. In Ref. [12], the authors have not considered the uplink real-time scheduling and resource allocation framework for the demand of 3GPP LTE-A networks. In Ref. [30], an OFDM multiplexing approach under multipath fading channels was proposed. In this methodology, the codes are permuted with the goal that the code bits contrasting with the hardened bits are doled out to subcarriers causing ceaseless piece goofs. This change can be considered as a kind of interleaving that can distinguishably upgrade the channel coding. Nevertheless, changing the codes dependent on the channel isn't always wanted practically speaking. That is on the grounds that it would result in a persistent change in the code of the correspondence channel type. This is viewed as a bulky, complex, and wasteful particularly for cloud-based applications. As a conclusion, the greater part of the previously mentioned methodologies accepts either the LTE-A or the IEEE 802.16 in their correspondence conventions. However, these measures have not been viewed as productive in planning. Subsequently, it is as yet an open research issue [22].

In this work, we aim at proposing an adaptive framework for dynamic channel management in smart cities. We propose a novel design for the mobile vehicular-cloud infrastructure that takes into account varying weather, road, and traffic conditions. This framework utilizes the latest channel coding techniques while utilizing existing cellular infrastructure to stream data, sound, and video with the least latency.

8.3 FRAMEWORK DESCRIPTION

In this section, we recommend a complete depiction for the proposed vehicular-cloud framework. We define our framework components as follows:

- *Management and Control Center (MCC)*: This component controls the velocity of the vehicle depending on the street conditions, traffic, and surrounding environment conditions.
- *Speed Limits Transmitters (SLT-x)*: This one is used to communicate the speed limit to the end-driver. The speed limits are transmitted as a wireless message. The MCC controls the speed of the vehicle and adapts it depending on the street, traffic, and weather conditions, given that enough Road-Side Units (RSUs) are located at predefined points on the road.
- *Speed Limits Receiver (SLR-x)*: This component takes the transmitted street upper limit and displays it clearly in the vehicle. It also is communicated to the driver's smartphone.
- *Vehicle Speed Sensor*: This is required to measure the speed of the vehicle accurately. The speed measurement system that already exists in the car can be used as well.
- *In-vehicle Microcontroller*: Its main task is to compare the actual speed with the road speed limit, which is received by the SLR-x. Based on the vehicle speed, the controller may generate an audio warning or may communicate such incidence to the MCC through the 5G network via the driver's smartphone.
- *5G Modem*: This modem is used to send a speed data to a central station which is monitored and operated by any other governmental or private entity.
- *Driver's Record Server (DRS)*: It stores information about the drive. The DRS is updated when the GSM modem sends a signal from the driver's car. The DRS can be accessible by third parties with the approval of the driver. Such third parties include parents, family members, and insurance companies.

8.3.1 FUNCTIONING OF THE VEHICULAR-CLOUD FRAMEWORK

Figure 8.1 shows a simplified schematic for the planned vehicular-cloud system, in normal conditions. The SLT-xs are connected to the MCC component through the public switched telephone network (PSTN) and the GSM/5G networks, which are usually connected to cloud data centers. The SLT-xs are controlled by the MCC, which recommends changing the speed limit depending on the road, weather, and traffic conditions. The MCC gets the deciding information from the patrol police, forecast stations, and driver's smart phones. The vehicle speed is continuously

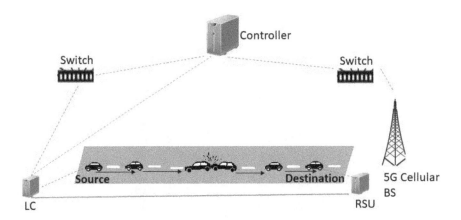

FIGURE 8.1   A schematic for the vehicular cloud in smart cities.

compared to the received speed limit which is also displayed to the driver. Hence, the driver will always know the upper limit of the street s/he is on. This is a better method than the conventional speed limit signs located on the side of the road. If a driver is driving below the speed limit, there is nothing to do except that it might be a congestion case. If the vehicle speed limit exceeds the received one, a warning signal will be generated to alert the driver that he exceeded the speed limit. If the driver does not respond within a given period, the In-vehicle microcontroller sends a note (violation) through the 5G modem to the drivers' record server (DRS). The violation can also be sent to the driver's smart phone. The violation can also be recorded using various network-positioning techniques. Other forms of violations such as tail tracking and red light tracking can be motored in the car which has other types of sensors. In such a case where a violation was recorded, the DRS can inform the driver instantaneously through his/her mobile/e-mail/mail, or by some other IEEE 802.16 standard means. We remark that this study focused on the voice over IP (VoIP) service as an example of the exchanged real-time contents over the vehicular cloud.

8.3.2   REGISTRATION PHASE

We assume that RSUs are located in different places along a given road. Local controllers (LCs) are static BSs in the city. Mobile controllers (MCs) are sensing platforms attached to vehicles. We use a minimum spanning tree (MST) algorithm to select an LC at each route. Each MC is considered as a node in the MST, whereas all RSUs are considered as terminal nodes, as depicted in Figure 8.2. Accordingly, this registration phase is summarized in the following four main steps:

Step 1: The MC first transmits a control message such as a Hello message to all in-range RSUs to determine the layout of the network as depicted in Figure 8.1.

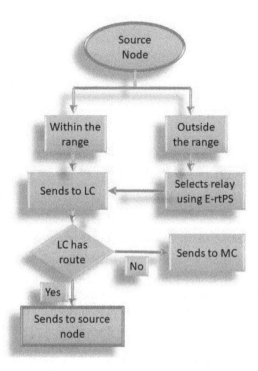

FIGURE 8.2 A workflow for the vehicular cloud.

Step 2: An RSU computes its delay and compares the delay with the neighboring RSUs in the same depth based on IEEE 802.16 standard means. The RSU with the lowest delay declares itself as the LC.

Step 3: The selected LC broadcasts a message to the MC and all RSUs in range, with the updated information about all the nodes, and these are vehicles and RSUs. All nodes contacted record the route to an MC in their flow table.

Step 4: The previous three steps are repeated at every step, and finally, the MC institutes a global layout and sends it to all LCs, RSUs, and vehicle networks.

We would like to remark that RSUs can calculate the number of Hello messages sent previously as the vehicle moves and updates each other and the nearby RSUs. This helps in improving the stability of the vehicular-cloud system by averting sparse network conditions and thus selecting an LC with the highest connectivity. We similarly calculate the hop count by looking at the number of RSUs a message has passed by (see Figure 8.2.).

At the end of this phase, the selected LCs create a localized global view at each depth from the MC. Therefore, different kinds of controllers are used to reduce the system burden form the single main controller and reduce overall overhead and delay.

8.3.3 Channel Coding Phase

We assume an IP real-time subsystem network (IRS-N)-based framework utilizing CQICH for demanding the bandwidth from the free wireless channel resources. We make use of the two reserved bits in the universal MAC header of IEEE 802.16. One of these bits, which is called the grant size (GS) bit, is responsible for informing the BS about the state of the voice transition. For example, when the voice connection of the MCC is on, the MCC ascribes the GS bit to *zero*. Moreover, the MCC imparts any changes in the voice transition using the traditional MAC header. So, the BS keeps informed of the changes happen to the voice state without causing any MAC overhead. Alternately, the uplink resources can be used to get the bandwidth request process while using the MCC and BS resources to monitor the GS bit transmission. The BS assignment considers the uplink resource to realize an enhanced *real-time batch Markovian arrival process (RT-BMAP)*. Accordingly, many performance problems, such as delay, medium access, bit/frame error rate, and resource management, arise while supporting the multimedia services in a smart city. We assume the reserved bits of IEEE 802.16 are used for this purpose. Furthermore, the MCC is used to privilege the SC bit. The IRS network uses the required bandwidth in order to hold the uplink usage and the frame of voice codec acquired by the network.

8.4 RESULTS AND DISCUSSIONS

In this section, we present the results of our planned framework. It involves hardware and software implementation parts. The main purpose of the hardware part is to construct a simple test bed for such projects/ideas. This test bed has been proved to be very helpful with the cost approximation, and it provides critical information about design challenges such as the delay and system throughput. This simple test bed is composed of three subsystems: the in-vehicle subsystem connected via cellular networks, the IP-based network (Internet), and the management and control center (MCC) represented by the network client.

We assume the IEEE 802.16 model is built in the MAC layer to use UGS and E-rtPS as an online resource for voice services. Also, we deploy real-time client (RC) and real-time server (RS) as agents to support the voice call creation and dissolution. These agents use the real-time transport protocol (RTP)/user datagram protocol (UDP) to control messages transmissions.

We remark here that we assumed ten cars while obtaining the results in this section for realistic verification purposes. The MCC is implemented via Ubuntu-PC (laptop) for experimental purposes. And the car-toys were equipped with Arduino boards attached to GPRS modules. To evaluate the performance of the assumed resource allocation framework in the central cell, we consider using 5 MHz LTE-A with 19 cells working at 2.0 GHz as a wireless cellular network. For each user, the packet arrival process follows the poison distribution with a mean arrival rate of 1 Kbits/sec. In Table 8.2, our stimulation parameters are presented.

Figure 8.3 depicts the use of an IRS core function in Ubuntu, consisting of Proxy Call Session Control Function (P-CSCF), Interrogating Call Session Control Function (I-CSCF), Servicing Call Session Control Function (S-CSCF), and Home Subscriber

TABLE 8.2

Simulation Parameters of IMS Networks

Parameters	Values
Number of cells	19
Bandwidth	5 MHz
Shadowing standard deviation	8 dB
Number of RBs	25
Path loss	$128.1 + 3.76 \log(R)$, R in km
R	500 m
Frequency	2 GHz
Uplink transmission power	24 bBm
Modulation scheme	QPSK, 16, 64 QAM
Proximity distance	10 m
Maximum transmission P_m	24 dBm
Threshold η	0.8
Channel model	200Tap

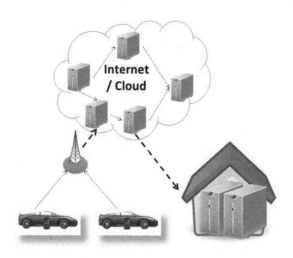

FIGURE 8.3   Experimental test bed for the vehicular cloud system of IEEE 802.16d/e.

Station (HSS) in order to monitor the quality of the online service. The system depicted in Figure 8.3 not only possesses the ability to allocate the network service resources but can also mask out roaming restrictions [23,24].

We use IRS core functions so as to transform voice connection to real-time service. The data can be transformed in serial or parallel mode to investigate the efficiency of E-rtPS with RT-BMAP compared to UGS and rtPS. In Ref. [25], the author said that through the knowledge of 3GPP, RTS provides a better service scalability to the devices connected; therefore, an IRS core system, with features of LTE-Advanced has been deployed, so as to provide better scalability and services. In this study, the core network is assimilated with HSS so as to analyze the stability of voice services across available networks. Moreover, the IRS core is embedded with its call session control function (CSCF) so as to generate a realistic test bed. Since the system is designed for real-time services, we gradually increase the voice connectivity in order to examine the system throughput and connectivity delay. Moreover, an emergency call is registered as unidentified caller to analyze session setup delay and throughput of real-time services. We perform an experiment to weigh the performance of SIP-based VoIP compared to IRS core using wireless connectivity and also to inspect the voice connectivity delay and throughput of the network. We use IRS clients and packet analyzers in order to establish a connection between VoIP call and service connection.

As depicted in Figure 8.4, the throughput of the proposed algorithm is much higher compared to UGS or rtPS. Moreover, Figure 8.5 shows that the delay of the proposed algorithm is lower at all critical points in comparison with the other services. It is imperative to note that getting the predefined delay of a service is important in the system of IEEE 802.16d/e [5] for the packet with delay violations.

Figure 8.6 depicts the bit error rate performance versus the energy per bit of the wireless channel codes using our proposed framework. From this figure, we can conclude that the achieved performance over a multi-way channel is almost as same as that obtained in Ref. [16]. The gain is also attained as a result of canceling the dispersion and fading effects. This can dramatically enhance the mobile communication channel performance in terms of reliability.

Meanwhile, Figure 8.7 depicts the frame error rate performance versus energy per bit of the wireless channel codes at varying block sizes. Obviously, as the block length increases, the performance gets improved. It's worth mentioning that the obtained reliability in terms of bit and/or frame error rate of the proposed framework makes the designed codes more appropriate over multipath fading channels. However, this might be associated with a slight channel loss at the maximum exchange rate. Therefore, quantifying the amount of data exchange rate incurred by the proposed framework is recommended.

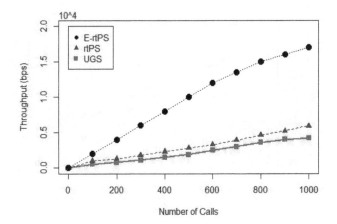

FIGURE 8.4 Throughput versus number of calls.

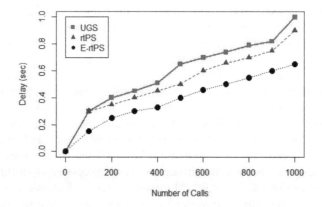

FIGURE 8.5   Average packet delay versus number of voice connections.

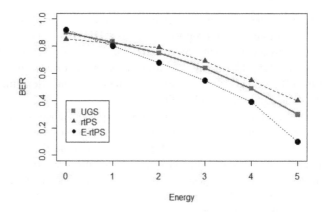

FIGURE 8.6   Bit error rate versus energy per bit.

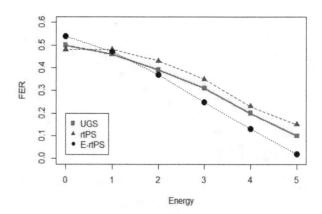

FIGURE 8.7   Frame error rate versus energy/power per bit.

8.5 CONCLUDING REMARKS

This chapter presents an adaptive framework for dynamic speed management in smart cities. The system makes use of the latest development in wireless communication and exploits the existing telecommunication infrastructures which have been used in data streaming, sound, and video to maximize the system adaptability and reduce the price. A key component to the planned framework is the dynamic medium access approach for real-time communications. Accordingly, this chapter proposes the Enhanced real-time polling system (E-rtPS) for the vehicular-cloud of IEEE 802.16. The proposed approach integrates RT-BMAP to analyze the throughput rate and average packet delay. It enables using the same coding design used for multipath selective channels while maintaining the same reliability performance in terms of delay and throughput.

In future work, data that are collected from the RSUs are going to be published to the cloud after preprocessing stage. The cloud will use machine learning and prediction techniques towards more efficient decisions. These decisions will be communicated back to relevant drivers on the road.

REFERENCES

1. R. Iqbal, T. A. Butt, M. O. Shafique, M. W. AbuTalib, and T. Umer, "Context-Aware Data-Driven Intelligent Framework for Fog Infrastructures in Internet of Vehicles." *IEEE Access,* vol. 6, pp. 58182–58194, 2018.
2. S. Ali, A. Ahmad, R. Iqbal, S. Saleem, and T. Umer "Joint RRH-Association, Sub-Channel Assignment and Power Allocation in Multi-Tier 5G C-Rans." *IEEE Access,* vol. 6, pp. 34393–34402, 2018.
3. X. Baró, S. Escalera, J. Vitria, O. Pujol, and P. Radeva, "Traffic Sign Recognition Using Evolutionary Adaboost Detection and Forest-ECOC Classification." *IEEE Transactions on Intelligent Transportation Systems*, vol. 10, pp. 113–126, Mar. 2009.
4. J. van de Beek, M. Sandell, and P. Börjesson, "ML Estimation of Timing and Frequency Offset in OFDM Systems." *IEEE Transactions on Signal Processing*, vol. 45, pp. 1800–1805, Jul. 1997.
5. F. Al-Turjman and A. Abdulsalam, "Smart-Grid and Solar Energy Harvesting in the IoT Era: An Overview." *Wiley's Concurrency and Computation: Practice and Experience,* 2018. DOI: 10.1002/cpe.4896.
6. S. Alabady, F. Al-Turjman, and S. Din, "A Novel Security Model for Cooperative Virtual Networks in the IoT Era." *Springer International Journal of Parallel Programming,* 2018. DOI: 10.1007/s10766-018-0580-z.
7. A. Klemm, C. Lindermann, and M. Lohmann, M, "Modelling IP Traffic Using the Batch Markovian Arrival Process." *Performance Evaluation*, vol. 54, no. 2, pp. 149–173, Oct. 2003.
8. F. Al-Turjman, H. Hassanein, W. Alsalih, and M. Ibnkahla, "Optimized Relay Placement for Wireless Sensor Networks Federation in Environmental Applications." *Wiley: Wireless Communication & Mobile Computing Journal*, vol. 11, no. 12, pp. 1677–1688, Dec. 2011.
9. D. Hui, S. Sandberg, Y. Blankenship, M. Andersson, and L. Grosjean, "Channel Coding in 5G New Radio: A Tutorial Overview and Performance Comparison with 4G LTE." *IEEE Vehicular Technology Magazine*, vol. 13, no. 4, pp. 60–69, Dec 2018.
10. F. Al-Turjman, "Fog-based Caching in Software-Defined Information-Centric Networks." *Elsevier Computers & Electrical Engineering Journal*, vol. 69, no. 1, pp. 54–67, 2018.

11. F. Al-Turjman and S. Alturjman, "Context-Sensitive Access in Industrial Internet of Things (IIoT) Healthcare Applications." *IEEE Transactions on Industrial Informatics*, vol. 14, no. 6, pp. 2736–2744, 2018.

12. D. Deebak, E. Ever, and F. Al-Turjman, "Analyzing Enhanced Real Time Uplink Scheduling Algorithm in 3GPP LTE-Advanced Networks Using Multimedia Systems." *Transactions on Emerging Telecommunications,* 2018. DOI: 10.1002/ett.3443.

13. A. Bravo-Santos, "Polar Codes for the Rayleigh Fading Channel." *IEEE Communications Letters*, vol. 17, no. 12, pp. 2352–2355, Dec 2013.

14. J. J. Boutros and E. Biglieri, "Polarization of Quasi-Static Fading Channels." in *Proceedings of IEEE International Symposium on Information Theory*, Jul. 2013, pp. 769–773.

15. F. Campioni, S. Choudhury and F. Al-Turjman, "Scheduling RFID Networks in the IoT and Smart Health Era." *Journal of Ambient Intelligence and Humanized Computing*, 2019. DOI: 10.1007/s12652-019-01221-5.

16. S. Choudhury and F. Al-Turjman, "Dominating Set Algorithms for Wireless Sensor Networks Survivability." *IEEE Access Journal*, vol. 6, no. 1, pp. 17527–17532, 2018.

17. P. Trifonov, "Design of Polar Codes for Rayleigh Fading Channel." in 2015 *International Symposium on Wireless Communication Systems*. IEEE, Aug 2015, pp. 331–335.

18. F. Al-Turjman and S. Alturjman, "Confidential Smart-Sensing Framework in the IoT Era." *The Springer Journal of Supercomputing*, vol. 74, no. 10, pp. 5187–5198, 2018.

19. F. Al-Turjman, M. Z. Hasan, and H. Al-Rizzo, "Task Scheduling in Cloud-based Survivability Applications Using Swarm Optimization in IoT." *Transactions on Emerging Telecommunications,* 2018. DOI: 10.1002/ett.3539.

20. I. Mehmood, A. Ullah, K. Muhammad, D. Deng, W. Meng, F. Al-Turjman, M. Sajjad, and V. Albuquerque, "Efficient Image Recognition and Retrieval on IoT-Assisted Energy-Constrained Platforms from Big Data Repositories." *IEEE Internet of Things,* 2019. DOI: 10.1109/JIOT.2019.2896151.

21. E. Arikan, "Channel Polarization: A Method for Constructing Capacity-Achieving Codes for Symmetric Binary-Input Memoryless Channels." *IEEE Transactions on Information Theory*, vol. 55, no. 7, pp. 3051–3073, Jul. 2009.

22. H. C. Hsieh and J. L. Chen, "Distributed Multi-Agent Scheme Support for Service Continuity in IMS-4G-Cloud Networks." *Computers and Electrical Engineering*, vol. 42, pp. 49–59, Feb. 2015.

23. 3GPP TS 23.167, IP Multimedia Subsystem (IMS) emergency sessions, Release 11, Sept. 2013.

24. F. Al-Turjman and S. Alturjman, "5G/IoT-Enabled UAVs for Multimedia Delivery in Industry-Oriented Applications." *Springer's Multimedia Tools and Applications Journal*, 2018. DOI: 10.1007/s11042-018-6288-7.

25. S. Alabady and F. Al-Turjman, "A Novel Approach for Error Detection and Correction for Efficient Energy in Wireless Networks." *Springer Multimedia Tools and Applications*, 2018. DOI: 10.1007/s11042-018-6282-0.

26. H. Lee, T. Kwon and D. H. Cho, "Extended-rtPS Algorithm for VoIP Services in IEEE 802.16 Systems." *Proceedings of the IEEE International Conference on Communications*, pp. 2060–2065, Jun. 2006.

27. F. Al-Turjman and H. Hassanein, "Towards Augmented Connectivity with Delay Constraints in WSN Federation." *International Journal of Ad Hoc and Ubiquitous Computing*, vol. 11, no. 2, pp. 97–108, 2012.

28. M. Patel, "SOS Uniform Resource Identifier (URI) Parameter for Marking of Session Initiation Protocol (SIP) Requests related to Emergency Services." draft-patel-ecrit-sos-parameter-07.txt (October 26, 2009)

29. M. Oda and T. Saba, "Polar Coding with Enhanced Channel Polarization under Frequency Selective Fading Channels." in *2018 IEEE International Conference on Communications (ICC)*, May 2018, pp. 1–6.

30. F. Al-Turjman and C. Altrjman, "Enhanced Medium Access for Traffic Management in Smart-cities' Vehicular-Cloud", *IEEE Intelligent Transportation Systems Magazine*, 2020. DOI: 10.1109/MITS.2019.2962144.

9 Smart Parking and the Grid

Fadi Al-Turjman and Arman Malekloo
Near East University

CONTENTS

9.1   INTRODUCTION

The idea of smart parking was introduced to solve the problem of parking space and parking management in megacities. With the increasing number of vehicles on roads and the limited number of parking spaces, the congestion of vehicles is inevitable. This congestion would lead to driver aggression as well as environmental pollution. These factors may worsen particularly during peak hours where the flow density is at its maximum, locating a vacant parking spot is near to impossible. A recent report by INRIX [1] shows that on average, a typical American driver spends 17 hours a year looking for a parking space. However, looking at a major city such as New York, this figure is much higher. According to the report, New York drivers spend 107 hours per year searching for parking spots. Taking into account the amount of fuel spent during this period, significant levels of the emissions and the harmful gases are expected to appear. Identifying these problems and trying to resolve them in a manner that is effective and at the same time sustainable is a challenging task.

In the context of a smart city ecosystem, inputs from elements such as vehicles, roads, and users have to be networked and analyzed together in order to provide the best service in fast and secure manners [2]. One of the reasons for marching towards a smart city ecosystem is to use the potential of existing technologies and infrastructures in providing the best utility to users and improving their future. With the help of Internet of Things (IoT) applications, mobility and transportation are considered to be the key influencing factors in sustaining our surrounding environments, especially those which utilize intelligent transportation system (ITS) [3].

One of the key components in ITS is the smart parking system (SPS), which relies significantly on analyzing and processing the real-time data gathered from vehicle detection sensors and the radio-frequency identification (RFID) systems that are placed in parking lots to report the absence and/or presence of a vehicle. These sensors have their strengths and weaknesses in certain areas where they are deployed. In addition, there might be issues in data anomalies and discrepancies where the collected information does not always conform to the initially expected pattern. This could potentially lead to a less reliable system. Moreover, security and privacy issues of the data transmitted and/or received must be carefully treated. Several factors such as communication and data encryptions must be well investigated in advance before implementing such systems. These subjects have to be seriously considered as the data collected from these sensors might be used in several critical scenarios such as the parking space prediction in emergencies and the path optimization in self-driven cars. Any vulnerability in these scenarios, no matter how significant/insignificant they are, can potentially lead to personal information leakage and increase the risk of security attacks. Considering the aforementioned aspects can significantly enhance the experience of both the parking lot operators (by maximizing their revenues) and the SPS users (by easily searching, booking, and paying in advance for their parking lot). Accordingly, any smart application in the current smart city ecosystem has to be context-aware and has to adapt dynamically to contiguous changes.

In Ref. [4], the authors presented the main motivations in carrying smart devices and the correlation between the user surrounding context and the used application. They focused on context-awareness in smart systems and space discovery paradigms. New generations of real-time monitoring in SPS have recently been discussed and gained attention as well. Vehicular ad-hoc networks (VANET), in which vehicles are communicating with other nearby vehicles as well as roadside units, play a key role in SPS while providing a real-time parking navigation service [4]. Authors have considered the development of a hybrid sensor and vehicular network for safe ITS applications. Another trend is the unmanned aerial vehicles (UAVs), which provide wireless connectivity in locations where the cellular range is limited or the existing infrastructure fails to operate. Equipping UAVs (or drones) with the vehicle detection sensors can ultimately solve many problems that the current deployments of smart parking systems are facing. Furthermore, with the emerging long-range low-power wide area networks (LPWANs) and the fifth-generation (5G) networks, several IoT services can be provisioned in SPS. In Ref. [5], the authors describe the IoT paradigm as a dynamic global network infrastructure with self-configuring capabilities based on standard and interoperable communication protocols. With the increasing number of connected devices day by day, the need for faster and more efficient wireless communication trend in parking systems is expected. This is where 5G can fill the gap [6]. However, interoperability issues related to both software and hardware aspects in these solutions can dramatically degrade the system performance. And thus, ubiquitous IoT solutions in SPS need to be open and able to integrate with other existing platforms.

9.1.1 The Scope of This Chapter

This survey aims to offer an insight into new parking paradigms in ITS. We look at different aspects of the SPS. First, we define and classify the existing smart parking attempts in order to provide the reader an idea about the concept of the SPS in general and its possible categories/alternatives. For a comprehensive study, we cover both hardware and software aspects in this system. In terms of hardware, we investigate all vehicular sensors that are currently in use via several use cases in terms of their strengths and weaknesses. Moreover, we list and overview the different physical communication technologies that are especially used in smart parking, while identifying the main elements influencing the parking system performability. Summary tables have been also added to offer a quick and rich overview about these sensors and communication technologies. On the other hand, relevant software systems and algorithms used to assure smooth and comfort quality of service (QoS) to the end user have been also investigated. We outlined and analyzed potential algorithms in parking prediction, path optimization, and assisting techniques in the literature. We also delved into the soft security issues and assessed the literature for countermeasures against the existing security attacks. For more uniform and open access solutions, we discussed the system interoperability from the soft privacy point of view in the context of smart parking. We further present new emerging applications and software systems handling common parking problems with the help of VANET paradigms, in addition to recommending a novel conceptual hybrid SPS.

In general, this survey aims to gather and zoom in critical design aspects that might be of interest to readers from the academia as well as the industry. It provides a multidisciplinary source for those who are interested in ecosystems and IoT-enabled smart cities. Also, it opens the door for further interesting research projects and implementations in the near future. For more readability, the used abbreviations in this survey along with their definitions are provided in Table 9.1.

9.1.2 Comparison of Other Surveys

There have been similar attempts and surveys in the literature on smart parking solutions with their own merits and limitations. In this subsection, we overview these attempts and highlight how it contrasts with our survey. For example, a very short survey on the parking lot reservations in cloud-based systems was discussed in Ref. [7]. The explanation for the mechanism of each reservation technique was briefly discussed with examples. A similar survey was provided in Ref. [8], however, without examples from the literature. The authors have slightly touched a few types of the vehicular sensors. Nevertheless, it was not comprehensive and did not cover all types of possible sensors and limitations. The inadequacy of the previous surveys was covered and better expressed with more examples and justifications in Ref. [9]. The authors presented some of the vehicular detection sensors with examples from the literature. However, it fails to demonstrate other aspects of the SPS such as the utilized communication protocols and/or software systems. With respect to communications, the authors in Ref. [10] explained the different implementation aspects of the wireless sensor communication protocols and later proposed an adaptive and

TABLE 9.1
List of Used Abbreviations

Abbreviation	Description	Abbreviation	Description
ABGS	Agent-based guiding system	MQTT	Message queuing telemetry transport
AI	Artificial intelligence	MSN	Mobile storage nodes
AMR	Anisotropic magneto-resistive	NAPS	Non-assisted parking search
APTS	Advanced public transport system	NB-IoT	Narrowband Internet of Things
AVI	Automated vehicle identification	OAPS	Opportunistically assisted parking search
CAPS	Centralized-assisted parking search	OBU	Onboard unit
CCTV	Closed-circuit television	O-DF	Open data format
COINS	Car park occupancy information system	O-MI	Open messaging interface
DSRC	Short-range communication	PGIS	Parking guidance and information system
ECC	Elliptic curve cryptography	PLRS	Parking lot recharge scheduling
EV	Electric vehicle	PRS	Parking reservation system
FCFS	First come first serve	QoL	Quality of life
FMCW	Frequency-modulated continuous wave	QoS	Quality of service
GAN	Generative adversarial network	RFID	Radio-frequency identification
GIS	Geographic information system	RSU	Roadside unit
GPS	Global Positioning System	SDN	Software-defined network
IIOT	Industrial Internet of Things	SPS	Smart parking system
ILD	Inductive loop detector	SVG	Scalable vector graphics
IoT	Internet of Things	TBIS	Transit-based information system
IoV	Internet of vehicles	UAV	Unmanned aerial vehicles
ITS	Intelligent transportation system	V2I	Vehicle to infrastructure
LDR	Light-dependent resistor	V2R	Vehicle to roadside
LoRaWAN	Long-range wide area network	V2V	Vehicle to vehicle
LPWAN	Low-power wide-area network	VANET	Vehicular ad-hoc networks
LTE	Long-term evolution	VIM	Vehicle in motion
MANET	Mobile ad-hoc network	VMS	Variable message sign
MAV	Microaerial vehicle	WSN	Wireless sensor network
ML	Machine learning		

self-organized protocol. Despite this, they did not consider the emerging trends in LPWAN communication protocols and mostly relied on sensors and RFID systems while conducting their research. Multi-agent, fuzzy, vision, and VANET-based smart parking methods were overviewed in Ref. [11]. The authors talked about the used software systems in general, but they did not go much into the hardware details. A related, albeit brief, summary of smart parking software systems and applications with different advantages and disadvantages was presented in Ref. [12]. However, like the other surveys, reasonable parts of the necessary information and hardware details were missing. Perhaps the most comparable survey to our work is given in Ref. [13]. This survey referred to reasonable aspects of the smart parking system such as information collection, dissemination, and deployment. However, it relies on outdated references and lacks the discussion of new enabling technologies related to drones and other emerging sensors in the field. Our survey can be considered one of the most comprehensive and up-to-date studies in comparison with the aforementioned surveys with additional insights and discussions about the new trends in SPS, which were briefly mentioned and/or totally ignored in the literature. In addition to covering the detailed hardware components, data interoperability, privacy, and security issues are discussed and assessed. Moreover, hybrid solutions, with more comprehensive analysis in terms of communication networks and recent innovative parking applications, are discussed in the smart city context. In Table 9.2, we summarize these differences in comparison with the aforementioned surveys and contributions in the literature.

The rest of this chapter is organized as follows. Section 9.2 classifies the existing smart parking systems into centralized versus distributed systems and discusses their main features. In Section 9.3, we overview and compare the current state of vehicular detection technologies while focusing on critical parameters such as scalability, accuracy, and sensitivity to weather conditions. Utilized sensors are also classified into active versus passive sensors. In Section 9.4, common design factors are listed under three main categories: soft, hard, and interoperability factors. Varying use cases have been presented and discussed in the context of both small-scale individual vehicular systems and large-scale implementations for smart city ecosystems. We also investigate the new generation of smart parking systems utilizing VANET and drone paradigms, in addition to recommending a new conceptual hybrid model for the smart parking system. In Section 9.6, we overview existing open research issues in smart parking systems ranging from the sensor limitations to the communication network capabilities, as well as the need for energy harvesting and data interoperability. Finally, we conclude our survey in Section 9.7.

9.2 SMART PARKING SYSTEMS AND CLASSIFICATIONS

Smart parking systems are categorized into various categories in which each of them has a different purpose and use different technologies in detecting vehicles. Smart parking systems benefit both the drivers and the operators. Drivers use the system to find the nearest parking spots, and the parking operators can utilize the system and the collected information to agree on better parking space patterns and a better pricing strategy. For example, since the demand for a parking lot is not stable, using a dynamic pricing approach, which takes into consideration the time and type of customers, can help the operators boost their revenues [14]. Smart parking enables several attractive services such as the smart payment/reservation, which can substantially enhance the experience of both drivers and operators. Moreover, the smart

TABLE 9.2
Summary of Related Surveys

References	Classification	Sensors	Communication Protocols	Software System	Security and Privacy	Interoperability and Data Exchange	New Applications	Open Issues
[7]	–	✓†	–	✓†	–	–	–	–
[8]	✓*	✓*	–	–	–	–	–	–
[9]	✓*	✓*	–	–	–	–	–	–
[10]	✓†	✓†	✓*	✓†	–	–	–	–
[11]	✓*	✓†	✓†	✓*	–	–	✓*	–
[12]	–	✓*	✓*	✓*	–	–	–	–
[13]	✓*		✓*	✓	✓†	–	✓*	✓
Our survey	✓	✓	✓	✓	✓	✓	✓*	✓

✓ Represents a comprehensive analysis.

✓* Represents a general analysis.

✓† Represent very little analysis.

– Means not considered.

parking system helps in preventing the unauthorized vehicular usage, as it increases the security measures on parking lots. Furthermore, SPS can play a significant role in providing a clean and green environment by minimizing the vehicle emissions via decremented delays in finding the vacant parking spot [15].

SPS architectures commonly consist of several layers based on their functionalities [8,16]. First, the sensing layer – it is the backbone of the SPS and is responsible for detecting the presence and/or absence of a vehicle in an area using different sensing technologies. These technologies are mostly comprised of receivers, transmitters, and anchors. Second, the network layer – it is the communication segment of the system, which is responsible for exchanging messages between transmitters/receivers and the anchors. Third, the middleware layer – it is the processing layer of any SPS in which intelligent and sophisticated algorithms are utilized to process the real-time data. It also acts as a data storage as well as the link between the end users requesting services from the lower layers. Finally, the application layer – it is the top layer in the system, which interfaces the SPS with clients (end users) requesting different services from different mobile and/or stationary information panels as depicted in Figure 9.1. These multilayered parking systems can be categorized into the following three types.

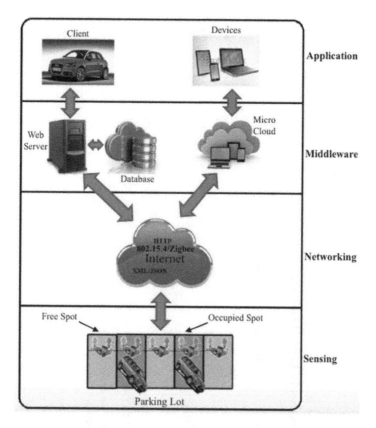

FIGURE 9.1 Smart parking system architecture. (Adapted from [8].)

9.2.1 CENTRALIZED-ASSISTED SMART PARKING SYSTEMS

In centralized smart parking systems, a single central server collects the necessary parking information and processes it to provide services such as the parking lot reservation, allocation, and/or driver guidance. The following sample systems are generally implemented as a centralized system.

9.2.1.1 Parking Guidance and Information System (PGIS)

PGIS, or also known as advanced public transport system (APTS), works by collecting parking information dynamically from loop detectors, ultrasonic, infrared, and microwave sensors in order to inform the drivers in real-time manners about the vacancy of the parking lot via an onboard guidance system or a variable message sign (VMS) [15,17]. PGIS consists of four major subsystems, namely, the information collection, processing, transmission, and distribution [18]. It can be implemented in both citywide and individual parking lots, where in both cases, drivers can easily follow and navigate to reach the vacant parking space [19]. In Ref. [17], a combination of PGIS and dedicated short-range communication (DSRC) is presented, where the DSRC-based PGIS provides a real-time, rapid ,and efficient way of guidance. However, concerns regarding the PGIS algorithm efficiency and the data safety, as well as the need for incorporating heterogeneous smart subsystems may cause severe issues in the implementation stage. A combined PGIS with a mobile phone terminal with the help of Global Positioning System (GPS) to locate and predict vacant spots and to guide drivers to the destination was presented in Ref. [18], whereas Shiue et al. utilized both GPS and 3G in Ref. [20]. Reliability of GPS and 3G connectivity in a multilevel parking lot is an issue, which may cause such systems to be impractical and ineffective. In Ref. [21], the authors proposed a PGIS in combination with ultrasonic sensors and WSNs. Meanwhile, Ref. [22] integrated RFID- and Zigbee-based solutions. These sensor-based solutions have their own disadvantages as we are going to explain later in this chapter. In general, all processing and decision-making processes in PGIS are performed at a central processor (server) [23].

9.2.1.2 Centralized-Assisted Parking Search (CAPS)

In this example, the first come first serve (FCFS) approach is adopted, where the first requester vehicle is guided towards a guaranteed vacant spot closest to the driver location. However, in this manner, other vehicles waiting in the queue are in continuous movement until the server can satisfy them. This brings the issue of uncooperativeness between drivers that can significantly degrade the performance in CAPS. Furthermore, the high maintenance cost and scalability are serious concerns in CAPS as well. In Ref. [24], the authors proposed a parking lot recharge scheduling (PLRS) system under this category for electric vehicles. They compared the performance of their approach against basic scheduling mechanisms such as FCFS and earliest deadline first (EDF). Their optimized version of FCFS and EDF outperforms the basic mechanisms with regard to maximizing revenue and the number of vehicles in the parking lot. In another example, edPAS, the abbreviation for event-driven parking allocation system, focuses on effective parking lot allocation based on certain events in the parking lot while dynamically updating the communicator [25]. This system utilized both FCFS- and priority (PR)-based allocation schemes.

9.2.1.3 Car Park Occupancy Information System (COINS)

COINS utilizes video sensor techniques based on one single source to detect the presence and/or absence of vehicles. The status is then reported on information panels which are strategically placed around the parking lot [26]. COINS is mainly dependent on four different technologies: (1) the counter-based technology, (2) the wired sensor-based technology, (3) the wireless sensor-based technology, and (4) the computer vision-based technology. Knowing that using the last technology can provide more accurate results about the exact status of the parking spot without deploying other sensors in each individual spot [26,27], in Ref. [26], COINS was developed and simulated in different environments with different parameters such as weather conditions and illumination fluctuations, which can add an extra layer of complexity to the system. Applications of COINS in a multilevel parking lot may not be that much effective like the other parking systems due to scalability and coverage issues.

9.2.1.4 Agent-Based Guiding System (ABGS)

ABGS simulates the behavior of each driver in a dynamic and complex environment explicitly. The agent in this system is capable of making decisions and defines the interaction between drivers and the parking system based on perceived facts from the driver and other varying aspects such as autonomy, proactivity, reactivity, adaptability, and social ability. For instance, SUSTAPARK was introduced in Ref. [28] to enhance the searching experience while locating the parking space in an urban area with an agent-based approach. The authors aimed at dividing the parking task into manageable sub-tasks that the computer agents could follow using Artificial Intelligence (AI) techniques. Another agent-based approach was introduced in Ref. [29], named PARKAGENT, based on ArcGIS with similar functionalities to SUSTAPARK. However, it considers the effects of the system entities' heterogeneity and the population distribution of drivers. In Ref. [30], an agent-based system was used as a negotiator to bargain over the parking fee with guidance capabilities to guide drivers to the optimal parking destination using the shortest path. It was relying on some perceived factors and performed interactions with other agents.

9.2.1.5 Automated Parking

Automated parking consists of computer-controlled mechanical systems which enable drivers to drive their vehicles into a designated bay, lock their vehicles, and allow the automated parking system to manage the rest of the job [9,15]. Stacking cars next to each other with very little space in between allows this system to work in an efficient way such that the maximum available space in the parking is utilized. The retrieval process of the vehicles is as easy as entering a predefined code or password. The process is fully automated, which adds an extra layer of security and safety to the whole system including both drivers and vehicles [9]. Although the initial cost of automated parking is very high, however, for the services that you receive, the price is very competitive. In fact, 50% saving compared to conventional modes of the parking in locations, where the parking fee is high and time-limited, is expected [27]. In these systems, employment of one or a combination of many services and sensors may be integrated to provide a fast, reliable, and a secure mode of parking with little or no interactions between the drivers and the system. A general concern regarding such a system is that a universal building code does not exist yet. In order to serve all customers, compatibility issues with the varying vehicles' models' have to be addressed and solved.

9.2.2 DISTRIBUTED-ASSISTED SMART PARKING SYSTEMS

In distributed smart parking systems, many services are connected and are controlled by a single server. This is well explained in vehicular networks where one vehicle can exchange information to one or more vehicles effectively creating a distributed network of vehicles. Another example is in the systems where information processing and dissemination is generally based on roadside infrastructure. The following examples are considered as distributed and opportunistic smart parking system in the literature.

9.2.2.1 Transit-Based Information System (TBIS)

TBIS is a park- and ride- based guidance system with similar functionalities to PGIS. It communicates with drivers through VMS to guide them towards a vacant parking spot. It also provides a real-time information about the public transportation schedules/ routes status, which enables drivers to pre-plan their journey more efficiently [9]. A field test in San Francisco [31] shows promising results for the effectiveness of TBIS. However, due to the initial capital cost, such a system should be implemented mainly in large-scale applications to recover the cost. Geographic Information System (GIS) is also another way for providing traffic information to users [32,33]. This system provides the minimum traveling time by optimizing the route/schedule of the functioning public transportation system in real-time manner. It enables web-based GIS systems to be implemented for the convenience of users in planning their trips.

9.2.2.2 Opportunistically Assisted Parking Search (OAPS)

Vehicles with IEEE 802.11x communication standard in ad-hoc mode can share information about the status and location of the parking spots. This enables drivers to make more knowledgeable decisions, while they are searching in the crowd. In this approach, drivers are guided towards the closest vacant parking space by analyzing timestamps and geographical addresses using GPS units, for example. Since OAPS dissemination service does not impose a global common knowledge about the status of parking spots, the outdated timestamps and infrequent updates could cause delays and intimidate the effectiveness of this approach [23]. Another issue could be the misbehavior of drivers when they enjoy the shared information from other drivers, but show selfishness in sharing theirs [34]. This can increase the distance between the destination and the parking spot, in addition to increasing the parking search time.

9.2.2.3   Mobile Storage Node-Opportunistically Assisted Parking Search (MSN-OAPS)

Instead of normal vehicle nodes, the inflow of information is channeled through mobile storage nodes (MSNs), which enable information sharing with other mobile nodes acting as a relay between vehicles. Similar issues in data dissemination can also be observed as the status of the parking spot changes overtime because the accuracy of disseminated data has a tendency to drop down as the number of relays increases. In Ref. [34], the authors suggested that MSN could improve the performance of OAPS. However, it does not have a significant effect, when a selfish driver as we afore-described uses it.

9.2.3   NON-ASSISTED PARKING SEARCH (NAPS)

In NAPS approach, there is no inflow of information from any vehicle/server. The parking decision is solely dependent on the driver's observation in the parking lot, or on a former experience considering the traffic flow and the time of the arrival in the parking lot. Drivers wander around a parking lot and check for empty spots in sequence until an empty one is found. This empty spot is then allocated to the driver who has reached the spot first [23,35]. Usually, it is performed with the minimum technology involvement. That is why we call it, "non-assisted" parking.

Scalability and cost design issues associated with the aforementioned categories of smart parking systems are mainly justified/claimed by the amount and type of the utilized sensors and/or enabling technologies. For example, in a large-scale smart parking application (e.g., in the multi-story shopping malls), where multiple sensors of one or more type(s) are required, the overall cost may exceed the cost of using the same parking system in a small-scale application. Therefore, as noted in Table 9.3, the cost of a few smart parking systems might be dependent on the scale and the scope of the targeted application. Table 9.3 tabulates the afore-overviewed parking systems based on their varying parameters and requirements.

9.2.4   USE CASES IN PRACTICE

In this subsection, we outline three common use cases of the smart parking system for a discussion that is more comprehensive. These use cases are named as follows: (1) the smart payment system (SPayS), (2) the parking reservation system (PRS), and (3) the E-parking system. A detailed analysis for these use cases is provided in the following subsections.

9.2.4.1   Smart Payment System (SPayS)

Conventional parking meters were always slow and inconvenient to use. Nevertheless, the smart payment system has been developed and integrated nowadays with IoT and advanced technologies that assure reliability and fast payment methods [15]. This system uses contactless, contact, and mobile modes to achieve its purpose. In contactless mode, smart cards and RFID technologies such as automated vehicle identification (AVI) tags are used. In contact mode, credit and debit cards are utilized. In mobile mode, mobile phone services are used to collect the

TABLE 9.3

Classified Parking Systems

Parking System Classification	Sensors	Broadcast	Guidance	Scalability	Coverage	E-Parking	Reservation	Cost (O&M Incl.)
PGIS	All	VMS	✓	*****	Citywide, local	✓	✓	Scale dependent
TBIS	All	VMS	✓	*****	Citywide, local	✓	✓	Scale dependent
CAPS	All	VMS	✓(FCFS)	***	Local	✓	✓	*****
OAPS	Vehicle	V2V, V2I, MSN	✓	**	Local	✓	–	***
NAPS	–	–	–	*	–	–	–	None
COINS	Video and image processing	Information panel	✓	**	Local	✓	Comparison of the hybrid and single-sensor smart parking systems	***
ABGS	All	Agent	✓	****	Citywide, local	✓	✓	Scale dependent
Automated parking	Limited	Information panel	–	–	Local	✓	✓	****

* Represents a rating system, one star (*) means *poor* and the five stars (*****) mean *excellent* in terms of their used metric (e.g., accuracy and scalability), whereas, for the cost metric, more stars mean more expensive.

✓ Indicates only applicability in certain cases (applications) as discussed in each subsection.

payment [8,15]. In Ref. [36], the authors proposed image processing technology in conjunction with SPS by utilizing RFID technology. This enables drivers to recall their parking spot, which contains the duration information required to calculate the parking fee. Internet-connected parking meters can also be used as a tool for the parking patterns determination and prediction, especially for the on-street parking where machine learning techniques are applied. However, the technical issue with SPS is the reliability and integrity of the system in case of wireless signal interception and/or routing protocol attacks, which could compromise confidential information [37].

9.2.4.2 Parking Reservation System (PRS)

PRS is a new concept in ITS, which allows drivers to secure a parking spot particularly in peak hours prior to or during their journey [38]. The objective of PRS is to either maximize the parking revenue or minimize the parking fee. This has been achieved by formulating and solving a min–max problem. Implementation of PRS requires several components, namely, the reservation information center, the communication system between the users and the PRS, the real-time monitoring system for the parking lot, and an estimation for the anticipated demands [38]. Drivers can later use a variety of communication services such as SMS, mobile phone, or web-based applications to make the reservation of a parking space. SMS-based reservation was implemented in Ref. [39], where the integration of micro-RTU (remote terminal unit) and the microcontroller, in addition to other safety features, makes it a smart solution in PRS. Such a system is also scalable and capable of handling multiple requests from drivers. CrowdPark system is another example of PRS proposed in Ref. [40], where the system works by crowdsourcing and rewarding to encourage drivers to use the system and report the parking vacancy. Malicious users and accuracy of the parking lot are a concern in these types of systems. However, in the case of CrowdPark, a 95% of a success rate has been reported in San Francisco downtown area. ParkBid crowdsourced approach was proposed in Ref. [41] where this system, unlike other crowdsourced applications, is based on a bidding process. This process provides numerous incentives about the parking spot information that enable urgent requests to be satisfied, in addition to reserving the closest parking spot.

9.2.4.3 E-Parking System

E-parking, as the name suggests, provides a system in which users can electronically obtain information about the current vacancy of the parking lots from other services and sensors. Moreover, it makes reservations and payments all in one go without leaving the vehicle and before entering the parking lot. The system can be accessed via mobile phones or web-based applications. In order to identify the vehicle making reservations, a confirmation code is sent to the user's email and/or mobile phone through SMS which then can be used to verify the identity of the vehicle [15]. Majority of the smart parking deployments that are introduced in this chapter are an example of E-parking where information about the parking vacancy can be achieved in advance. ParkingGain [42] is another example of the E-parking, where the authors presented a smart parking approach by integrating the OBU installed on vehicles to locate and reserve their desired parking spots. Moreover, their system offered the value-added services to subsidize part of the running cost and create business opportunities as well.

9.3 SENSORS OVERVIEW

In this section, several sensing and vehicle detection technologies are discussed. They are the integrated part of information sensing in smart parking system. In order to choose the best alternative for each application, several design factors should be taken into account, knowing that these sensors can vary in their strengths and weaknesses. Vehicle detection sensors are mainly categorized into two types: intrusive (or pavement invasive) and non-intrusive (or non-pavement invasive) [43]. Table 9.4 summarizes the list of sensors that have been typically used in smart parking systems. In addition, it overviews numerous parameters that can influence the choice of the sensor type to be used in a specific smart parking application. The sensed data accuracy and complexity levels, as well as the sensor maintenance operations, can typically influence the cost associated with these sensors. Moreover, depending on the installation requirement, the overall cost may rise. As in the case of piezoelectric sensors,

TABLE 9.4
Comparison of Different Parking Sensors

Sensor Type	Intrusive	Count	Speed Detection	Multi-Lane Detection	Weather Sensitive	Accuracy	Cost
Passive IR	–	✓	–	–	✓	**	**
Active IR	✓	✓	✓	✓	–	**	**
Ultrasonic	–	✓	–	–	✓	***	*
Inductive loop	✓	✓	✓	–	✓	****	**
Magnetometer	✓	✓	✓	–	–	****	*
Piezoelectric	✓	✓	✓	–	✓	****	***
Pneumatic road tube	✓	✓	✓	–	–	****	****
WIM	✓	–	–	✓	–	**	****
Microwave	–	✓	✓	✓	–	***	***
CCTV	–	✓	✓	✓	✓	****	****
RFID	–	–	–	–	–	**	*
LDR	–	✓	–	–	✓	**	*
Acoustic	–	✓	✓	✓	✓	*	**

* Represents a rating system, one star (*) means *poor* and the five stars (*****) mean *excellent* in terms of their used metric (e.g., accuracy and scalability), whereas, for the cost metric, more stars mean more expensive.

✓ Indicates only applicability in certain cases (applications) as discussed in each subsection.

accuracy represents additional cost on top of the installation difficulty. Therefore, the best sensor is the one that can best fill the need of the parking owner, while maintaining the lowest cost without sacrificing accuracy and comfort levels. In the following, these sensors are discussed and classified into active versus passive sensors, which are discussed next.

9.3.1 ACTIVE SENSORS

Active sensors are those which need an external power source to operate and perform their task. In the following, we overview the most common parking sensors which are classified as active ones.

9.3.1.1 Active Infrared Sensor

Active infrared sensors are configured to emit infrared radiation and sense the amount that is reflected from the object. And hence, it can detect empty spaces in a parking lot. While these sensors can accurately take advantage of multilane roads (active) and discover the exact location and speed of a vehicle on a multilevel parking lot, they are prone to weather conditions such as heavy rain and dense fog, and very sensitive to the sun [44].

9.3.1.2 Ultrasonic Sensors

Ultrasonic sensors work in the same way as infrared sensors do, but they use sound waves as opposed to light. It transmits sound energy with frequencies between 25 and 50 kHz, and upon the reflection from the vehicle body, it can detect the status of the parking lot [45]. Similarly, it provides other useful information such as the speed of the vehicle and the number of vehicles in a given distance [46]. Like IR sensors, they are sensitive to temperature and environment. However, because of its simplicity in installation and low the investment cost, it is widely used in smart parking applications to identify vacant parking spaces [45,47].

9.3.1.3 CCTV and Image Processing

The closed-circuit television (CCTV) technology combined with image processing techniques can be effectively used in many parking lots to determine the presence/absence of a vehicle. The stream of video captured from a camera is transmitted to a computer where it is digitized and can be analyzed frame by frame using sophisticated image processing techniques. Since the CCTV is already in use at several parking lots for surveillance purposes, the implementation of these systems is much easier, knowing that a single camera can analyze more than one parking spot simultaneously which makes it even more cost-effective in wide-area implementations [8,47,48]. However, it is constrained in deployment to fields with an open view and without obstacles. Moreover, weather conditions can significantly affect the performance of these systems sometimes [38,45,48].

9.3.1.4 Vehicle License Recognition

In conjunction with CCTV and image processing units, plates of the ingress and egress vehicles can be captured and analyzed to give an estimation of the arriving/

departing vehicles in real-time manners. Furthermore, continuous monitoring of vehicles movement in the parking lot until they reach the pre-designated parking spaces, which have been allocated to them, is a desirable feature in these systems [38]. This also enables smart payment to be deployed which allows the drivers to exit the parking bay without any delay since the required information is already forwarded to the automated gate controller. However, bad weather conditions can disrupt the functionality of plate recognition systems. Additionally, privacy concerns regarding storing the information of vehicles on a public database are another flaw in these systems.

9.3.2 PASSIVE SENSORS

Passive sensors are those, which rely on detecting and responding to inputs from the surrounding physical environment without the need for dedicated power supply units. In the following, common examples of these types of sensors are overviewed.

9.3.2.1 Passive Infrared Sensor

Passive infrared sensors work by detecting the temperature difference between an object and the surrounding environment [49]. In contrast to active infrared sensors that emit IR waves in a predefined pulse rate, passive sensors identify the vacancy of a parking space by measuring the temperature difference in the form of thermal energy emitted by the vehicle and/or the road. Thus, it is only triggered when a vehicle is in the vicinity of the detection zone of the sensor. Unlike the other sensors' types, passive infrared sensors do not need to be anchored or tunneled on the ground or the wall. However, these sensors are mounted to the ceiling in the parking lot [50]. These sensors are prone to weather conditions effects, which can degrade the parking system performance sometimes.

9.3.2.2 LDR Sensor

Light-dependent resistor sensor or LDR in short, detects changes in luminous intensity. By assigning a primary source of light, such as the sun, and a secondary source, such as the other surrounding light, the vehicle creates a shadow that causes the light sensor to detect luminous intensity changes. This indicates the presence of the vehicle in the parking spot. Weather conditions such as rain and fog, and the change in the angle where the sunlight is assumed to arrive can affect the performance of these kinds of sensors and can lower the detection accuracy [51]. As for their installations, these sensors are usually deployed on the ground in the center of the parking spot.

9.3.2.3 Inductive Loop Detector

Inductive loop detector (ILD) is normally installed on road pavements in circular or rectangular shapes [47]. They consist of several wire loops, where the electric current is passing through and generates an electromagnetic field. This field inductance excites a frequency between 10 and 50 kHz in the presence and/or passing of a vehicle over the sensor. It actually causes a reduction in the inductance that leads the electronic unit to oscillate with higher frequency which ultimately sends a pulse to the controller indicating the passing/presence of the vehicle [8,45,47]. However,

ILD sensors necessitate road resurfacing, multiple detectors for better accuracy, high sensitivity to traffic load, and temperature variations, in addition to the traffic disruption in case of maintenance, are major drawbacks for this category of parking sensors [45].

9.3.2.4 Piezoelectric Sensor

Piezoelectric sensors detect the mechanical stress that is induced by the pressure/ vibration of the vehicle when it passes over these sensors. This pressure/vibration is converted into electrical charges that generate a voltage difference proportional to the weight of the vehicle. Thus, the measured voltage can be used in classifying the sensed vehicle in terms of its weight and axle configuration. Such sensors are usually susceptible to high levels of pressure and temperature. Their accuracy is highly dependent on the type of the material used and the way it was cut. For more accurate measurements, multiple of these sensors should be used simultaneously [9,45].

9.3.2.5 Pneumatic Road Tube

Pneumatic road tube encompasses air pressure sensor at one end of the tube, whereas the other end is sealed to prevent the air leakage. Once the vehicle crosses this tube, the sensor sends a burst of air pressure along the tube. This operation results in triggering an electrical switch to be closed and produces an electric signal in order to recognize the vehicle presence. A software analyzer can then identify the vehicle type from the associated weight and axle configuration. These sensors are cost-effective and offer quick/simple installation. Nevertheless, this type of sensors can lead to inaccurate axle counting in case of the long vehicles' passing (e.g., buses and/ or trucks). This results in less reliable parking vacancy information [45].

9.3.2.6 Magnetometer

Magnetometer functions in the same way as loop detectors. It senses the changes in the earth magnetic field that is caused by metallic objects, such as vehicles, passing over these sensors. The cause for such distortion is that the magnetic field can easily flow through ferrous metals in contrast to the air [52]. There are two types of magnetometers: the single axis and the double/triple axis magnetometer. Accuracy of detecting vehicles using the second type is much higher due to the fact that it uses two/three axes while identifying the presence of a vehicle. Nonetheless, both types are reliable and resist undesired weather conditions. Meanwhile, lane closure, pavement cut, and in some cases short-range detection and inability to detect stopped vehicles are considered as shortcomings in this category [45].

9.3.2.7 Vehicle-In-Motion Sensors

Vehicle-in-motion (VIM) sensors can precisely determine the weight of a vehicle and the portion of the weight distributed on the body axles. Data gathered from these sensors are extremely useful and heavily used by highway planners and designers as well as law enforcement procedures. There are four distinct technologies used in VIM: load cell, piezoelectric, bending plate, and capacitance mat [45,48]. Load cell uses a hydraulic fluid that triggers a pressure transducer to transmit the weight information. Despite its high initial investment, load cells are by far the most accurate

VIM systems. Piezoelectric VIM system detects voltage variations per the applied/ experienced pressure on the sensor. Such a system consists of at least one piezoelectric sensor and two ILDs. Piezoelectric sensors are among the least expensive sensors in the market. However, their accuracy in vehicle detection is lesser than load cells and bending plate VIM systems. Bending plate uses strain gauges to record the strain or the change in length when vehicles are passing over it. The static load of vehicles is then measured by dynamic load and calibration parameters such as the speed of vehicle and pavement characteristics. Bending plate VIM system can be used for traffic data collection. It also cost less than the load cell system, but its accuracy is not at the same level. Lastly, the capacitance mat system consists of two or more sheet plates, which carry equal but opposite charges. When a vehicle passes on top of these plates, the distance between the plates becomes shorter and the capacitance increases. The changes in the capacitance reflect the axle weight. The advantage of using this system is that it can easily operate in a multilane road. However, it suffers high initial investment costs.

9.3.2.8 Microwave Radar

Microwave radar generally transmits frequencies between 1 and 50 GHz by the help of an antenna, which can detect vehicles from the reflected frequency. Two types of microwave radar are used in this sector: Doppler microwave detectors (DMDs) and frequency-modulated continuous waves (FMCWs) [45,48]. In the former type, if the source and the listener were close to each other, the listener would perceive a lower frequency. Contrary, if they were moving apart from each other, the frequency would get higher. If the source was not moving, Doppler shift would not happen. In this case, continuous range of frequencies is transmitted, and the detector can then measure the distance to the vehicle and indicates its presence. Microwave radars are more effective under harsh weather conditions. It can also measure the speed of the vehicle and conduct data collections on multilane traffic flow. However, measuring the speed using Doppler detectors requires additional sensor types such as the aforementioned ones to collaborate in accomplishing this task.

9.3.2.9 RFID

RFID can be used for vehicle detection as well in the parking lot. RFID units (readers and tags) consist of a transceiver, transponder, and antenna. RFID tags or transponders units with its unique ID can be read via a transponder antenna [47]. It can be placed inside the vehicle to be identified by the reader antenna that is placed in the parking lot. The reader reads the tag and changes the parking spot status to be occupied. With this system, the delay can be minimized and the flow of the traffic in the parking lot becomes smoother. Due to the range limitation in RFID systems, the distances between the reader and the tag is of utmost importance [53]. It is worth remarking that RFID in hybrid systems has shown more reliability than the stand-alone RFID system [54].

9.3.2.10 Acoustic Sensor

Acoustic sensors can detect the sound energy produced by the vehicular traffic or the interaction of tires with the road. In the detection zone of the sensor, a single processor computer can detect and signal the presence of a vehicle from the gener-ated noises. Similarly, in the drop of the sound level, the presence of the vehicle signal is terminated. Acoustic sensor can function on rainy days and can also operate on multiple lanes. However, cold weather conditions and slow vehicles can degrade the accuracy of such sensors [45,47]. To overcome this limitation, machine-learning techniques have been incorporated with these sensors for better performance.

9.4 DESIGN FACTORS

In this section, we look at the design factors that influence the performance of the smart parking systems in practice. We divide these factors into three main categories. First, we overview soft design factors that deal with software aspects of the system. Several soft solutions that have been found in the literature have been presented and correlated. Second, the experienced hardware issues and critical design aspects related to utilized sensors and communication networks in smart parking systems have been intensively studied and discussed as well. Moreover, we investigate the system components interoperability and data exchange with more focus on relevant smart city applications. This category is unique in nature and has been rarely overlooked in the literature, although it can make a significant effect in terms of the system performability and cost. We believe that the discussions in this section can be beneficial to all relevant smart solutions in a smart city ecosystem. Proposed examples and tabulated conclusions can considerably assist those who are interested in this area to quickly grab the fundamental information they seek.

9.4.1 Soft Design Factors

This category of the relevant design factors deals with software aspects of the park-ing system as well as the processing of data collected via the aforementioned sen-sors. We also discuss under this category the potential influences of the experienced privacy and security aspects from the collected data.

9.4.1.1 Software Systems in Smart Parking

Software systems play a key role in smart parking applications. They are used in managing the collected data from the sensors and then analyzing it efficiently. The performed soft analytics are based on algorithms that vary in intricacy depending on the scale and the complexity of the parking application. Furthermore, using the collected data and applying certain machine learning methods, these analytics can be used to predict parking vacancy and optimize the selected vehicle route. This enables the parking operators to efficiently manage their parking lots, in addition to maximizing their revenues. In this section, the software aspect of smart parking systems and some interesting ideas such as path scheduling, path optimization, prediction, and parking assignment have been intensively discussed.

Managing the information gathered from multiple sensors on a multilevel parking lot requires a robust software system that can manage, monitor, and analyze the data in an efficient manner. Many of the existing smart parking applications are using a centralized server to store and manage their data. After performing data analysis, several services such as the parking reservation and guidance can be implemented. For example, in Ref. [55], the authors presented a smart parking IoT and a cloud-based system using real-time information. Services such as the parking lot payments, reservation, and confirmation are all processed via a mobile application. In case of overshooting the parking time, the system offers an automated extension for the parking lot reservation and failing to do so will impose certain fees against the driver. A context-aware smart parking system based on collaborative work between smart servers, smart objects, and smart mobile devices was proposed in Ref. [56]. The smart servers collect/process the city-context information and all other related information regarding the parking status and registered users. It relays the information to distributed smart objects in the surrounding environment to make changes in the availability of the parking space. It finally displays the results via a GUI on the smart device, where the user can search, book, and pay for the parking lot.

Guidance software systems are getting smarter nowadays. Unlike the old PGIS, several design factors have been taken into consideration. In Ref. [57], the authors applied Stackelberg game theory to the PGIS in order to model the experienced dynamic changes in drivers' behaviors. They aim at balancing the parking revenue against the average time required in searching for a parking spot. A utility-based guidance approach for parking in a shopping mall was discussed in Ref. [58]. This guidance approach uses an improved A* shortest-path algorithm that generates the optimal vehicle route based on six different factors associated with the user preference and the parking utility. MQTT guidance protocol was considered in Ref. [59] in order to share simultaneously the real-time parking lot vacancy information with at least 1000 users. Their web application in JavaScript presents the information on the layout of the shopping mall drawn by scalable vector graphics (SVG). Moreover, an Internet of Vehicle (IoV)-based guidance system was proposed in Ref. [60]. Onboard hardware in vehicles enables IOV to interact with everything around (vehicles, pedestrian, and sensors) which then can be used with the optimal path and parking algorithms embedded in the system to guide the drivers to the nearest parking spot. Continuous license plate recognition using video with gray level changes (GLC) algorithm and Dijkstra algorithm for locating and guiding to the nearest parking spot was experimented in Ref. [61]. The combination of license plate and GLC detection ensures that the system is viable in case of passing pedestrians on the parking spot or when they cover the license plates. The use of GPS in smartphones with the combination of a genetic algorithm to locate and navigate to the closest parking lot was developed in Ref. [62]. The authors presented their solution and were able to obtain accurate results in several case studies. Instead of using exploration algorithms for locating the nearest parking space, a learning mechanism was used in Ref. [63]. The authors used reinforced learning algorithms in conjunction with Monte Carlo approach to minimize the expected time to find a parking place in an urban area. They compared their algorithms against the tree evaluation and random methods. It was deduced that their algorithms are less complex and more efficient. In Ref. [64], the authors propose a cache replacement approach for Fog applications in software-defined networks (SDNs). This approach depends on three functional factors in SDNs. These three factors are age of data based on the periodic request, the popularity of on-demand requests, and the duration for which the sensor node is required to operate in active mode to capture the sensed readings. These factors are considered together to assign a value to the cached data in a software-defined network in order to retain the most valuable information in the cache for longer time. The higher the value, the longer the duration for which the data will be retained in the cache. This replacement strategy provides significant availability for the most valuable and difficult to sense data in the SDNs. Decisions based on when and where to park is largely based on driver's observations. Many factors such as accessibility, fee, and availability of the parking influence the decision of the drivers. On the other hand, decisions based on experience when drivers locate a free parking space with or without prior information of the availability always tend to congestion of that particular space, increase the searching time, and cause long queues. However, if the availability of the parking space can be predicted and disseminated in time, the driver's experience in locating the most suitable location can be enhanced. Moreover, the prediction of the parking availability offers the parking operators short- and long-term system checks that ultimately enables them to take preventative decisions in case of any system failure.

Nowadays, prediction is as easy as to collect data from the sensors and apply algorithms. Previously, prediction was based on historical data and surveys about the parking vacancy. In Ref. [65], data collected from off-street parking garages was used to create a short-term model that could forecast the changing characteristic of the parking space using the wavelet neural network method. The authors compared their research method with the largest Lyapunov exponents in terms of accuracy, efficiency, and robustness. Although the model was successfully tested, other important criteria such as driver's behavior and/or environmental characteristics were not taken into consideration. In Ref. [66], the authors proposed a real-time availability forecast (RAF) algorithm based on drivers' preferences and the availability of parking that are iteratively allocated using an aggregated approach. Their algorithm is updated upon each vehicle arrival and departure to predict the dynamic capacity and the parking availability. Their test simulation in a parking facility in Barcelona showed promising results with minor errors. The results were then compared with the numerical method, where they observed no significant difference between the two approaches. Prediction of the parking space availability of Ubike system in Taipei City was also experimented in Ref. [67]. The authors used regression-based models (mainly linear regression and support vector regression) to forecast the number of bicycles in Ubike stations. Their model, due to a constraint that bicycles are only circulating around the station, was different from a parking system used for cars. However, a similar approach can be used for the car-park facilities. A prediction mechanism based on three features' sets with three different algorithms for comparison, namely, regression tree, support vector regression, and neural networks were developed in Ref. [64]. With the datasets provided from San Francisco and Melbourne parking facility and based on their model, it was concluded that the regression tree is the least computationally intensive algorithm.

The parking search-space optimization is another part of the software system in smart parking that utilizes the collected information to minimize the searching time and to maximize the number of parking spaces in a given parking lot. In Ref. [68], the authors showed that their adaptive multi-criteria optimization model can effectively reduce the search time by 70% in an urban area. Multi-criteria such as the walking distance, price, and driving time were set based on the drivers' preferences. It was presented by a utility function with the objective to maximize the expected utility. In a study conducted in University of Akron [69], the authors used direct random search method to perform their optimization model based on mainly different classroom assignment options. However, other factors such as the parking search behavior, arrival and departure distributions, and locations of different buildings and parking facilities were also taken into account. Their model managed to reduce the parking search time by about 20%. A cooperative parking search among the vehicles searching for a parking space using V2V and V2I technologies was studied in Ref. [70]. The authors found that when vehicles search cooperatively, a search time reduction up to 30% can be achieved. They also concluded that drivers would benefit more, if drivers can exchange information before and after reaching their destination. An intelligent hybrid model for optimizing the parking space search based on Tabu metaphor and rough set-based approach was introduced in Ref. [71]. Tabu search was used as a complement to other heuristic algorithms, whereas the rough set was used as a tool to manage the noisy and incomplete data.

9.4.1.2   Privacy and Security in Smart Parking

It is estimated that the number of connected IoT devices would reach 20 billion by 2023 [72]. These devices are in constant communication and a large number of data that is being processed, aggregated, and shared with users can be intercepted and used for nefarious intentions [73]. The most crucial part of any smart application is to assure that the network supports end-to-end encryption and authentication. In the case of interconnected IoT services and devices, any vulnerability, no matter the size is, can interrupt one side of the system and cascade it to the rest of the system [74]. In the following, we list a few key things to be considered while securing any SPS:

a. Data collection, which if it has been limited to a certain extent, could greatly help to mitigate any risks. For example, storing a large amount of data can elevate security breaches, and the collection of a huge amount of personal data may be used in a way that it is out of the scope of consumer's expectations from the system.

b. Information sharing, which requires lots of optimization and analysis in any smart-based application. In this regards, service providers and technology partners should come into an agreement for secure data handling and used techniques to ensure user's privacy such as de-identification.

c. Reliability of servers, encryptions, and digital signatures are also as important as the risk management protocols and the physical security of the system.

d. Human errors/mistakes, intentional or unintentional, can also elevate security risks. Therefore, policies and procedures for training sessions among the SPS users are required to mitigate oversight issues.

e. Finally, transparency of any smart system assures the integrity of such a system, which offers accountability and clear policies in regard to data security and privacy.

In addition to the aforementioned generic considerations, a good smart parking application requires a secure end-to-end communication between the end user and the server. Since the majority of the smart parking solutions are established based on either web or mobile applications, it requires users in such systems to enter personal information such as their home/business address. Since these systems also keep track of the history of transactions including the credit card information, they are also considered as critical aspects in data privacy and security of the existing smart parking systems [73]. P-SPAN, or privacy-preserving smart parking navigation system, was developed in Ref. [75]. Its navigation system for locating and guiding drivers to a vacant parking spot using Bloom filter and vehicular communications using private mechanisms has shown to be an effective smart parking system with low computational and communication overhead. Another VANET-based approach similar to the previous study with privacy preserving in mind was discussed in Ref. [76]. Their system provided a secure navigation protocol with one-time credentials. Some communication protocols lack data encryptions or require high computational resources in order to function securely. However, in Ref. [73], the authors used elliptic curve cryptography (ECC) with LPWAN protocol as an alternative to other cryptography techniques in devices where there exist hardware limitations. The aim of the work performed in Ref. [77] was to design a typical network security model for cooperative virtual networks in the IoT era. This paper presented and discussed network security vulnerabilities, threats, attacks, and risks in switches, firewalls, and routers, in addition to a policy to mitigate those risks. It also provided fundamentals of secure networking system including firewall, router, AAA server, and VLAN technology. It presented a novel security model to defend the network from internal/external attacks and threats in the IoT era. In Ref. [78], the authors proposed a context-sensitive seamless identity provisioning (CSIP) framework for the IIoT (Industrial Internet of Things). CSIP proposes a secure mutual authentication approach using hash and global assertion value to prove that the proposed mechanism can achieve major security goals in a short time period. Furthermore, in Ref. [77], the authors proposed a solution for secure data collection. They used a repository of sensing information acting as a sink for the collected data and a mirror of reservation database, which is synchronized with the repository. In this fashion, drivers are the only elements that can access the mirror database in order to make payments, check the vacancy of parking lots, and perform reservation via mobile devices.

9.4.2   HARD DESIGN FACTORS

The communication networks surrounding the smart parking systems, from the legacy to LPWAN communication protocols, for small- and large-scale applications have been addressed in this subsection from the hardware perspectives. In addition, we discuss the influences of the experienced sensors errors on the designed parking system.

9.4.2.1   Communication Networks

The employed IoT sensors in smart parking systems vary in terms of the used communication protocols. However, they all can be categorized into long-range LPWANs or short-range wireless networks. LPWAN has been integrated with existing cellular technology to avoid any further infrastructure need [44]. The work performed in Ref. [40], for example, provides an overview for deployment strategies of femtocells that can support several smart applications in the IoT era. In addition, it presents major LPWAN standards such as LoRaWAN, Sigfox, Weightless (SIG), Ingenu, LTE-M, and NB-IoT. On the other hand, short-range communication technologies such as Bluetooth, Wi-Fi, and Zigbee have been used for short-distance communication in the SPS. A comparison of range versus bandwidth for both modern and legacy communication protocols is shown in Figure 9.2. Moreover, Table 9.5 summarizes the technical parameters for both short- and long-range methods of communications [78–85]. Libelium,[1] a WSN platform provider, has used both LoRaWAN and Sigfox in their Plug and Sense platform, which uses magnetic sensors to detect vehicles in parking spots. Huawei[2] solution for smart parking has resulted in 80% energy reduction in their Czech Republic trial. ZTE[3] trials in China have claimed 12% and 43% reduction in congestion and time spent searching free spots, respectively. These are some of the examples of NB-IoT-based smart parking solutions. Moreover, China Unicom Shanghai smart parking, which is developed by Huawei, uses 4.5G LTE-M commination protocol in their parking network. Nwave[4] and Telensa[5] solutions use the Weightless N protocol along with magnetic sensors for the vehicles' detection. A deployment of the NB-IoT and third-party payment-based smart parking system was studied in Ref. [86]. The authors proposed a cloud and mobile application platform

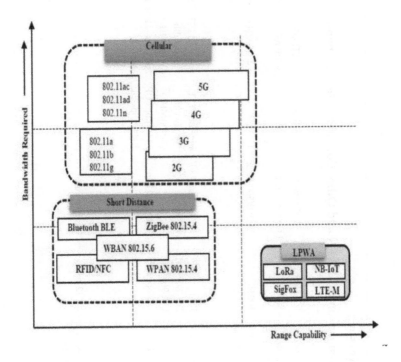

FIGURE 9.2 Comparison of range and bandwidth of LPWAN and other protocols [52].

TABLE 9.5

Comparison of LPWAN and Other Communication Technologies

Protocol	Sigfox	LoRaWAN	NB-IoT	LTE-M	Weightless W/N/P	Ingenu	Wi-Fi	Zigbee	Bluetooth/BLE
Standard/Mac layer	Sigfox	LoRa™ Alliance	3GPP rel. 8 and 13	3GPP rel. 13	Weightless Sig	IEEE 802.15.4k	IEEE 802.11b/g/n	IEEE 802.15.4	IEEE 802.15.1
Spectrum bandwidth	100 Hz	125/250/500 KHz	180 KHz	1.4–20 MHz	5 MHz (W) 200 Hz (N) 12.5 KHz (P)	1 MHz	80 MHz (2 antennas) 20 MHz (1 antenna)	868/915 MHz	79 channels 1 MHz 40 channels 2 MHz (BLE)
Frequency Band	ISM EU: 868 MHz US: 902 MHz	ISM 433/868/780/915 MHz	Licensed LTE bandwidth (7–900 MHz)	Licensed LTE bandwidth (7–900 MHz)	Sub GHz	ISM 2.4 GHz	2.4/5 GHz	2.4 GHz	2.4 GHz
Data Rate	100 bps UP 600 bps DL	290 bps–50 kbps	234.7 kbps UP 204.8 kbps DL	200 kbps – 1 Mbps	1 kbps – 10 mbps (W) 30–100 kbps (N) 200 bps–100 kbps (P)	78 kbps UP 19.5 kbps DL	11 Mbps (b) 54 Mbps (g) 1 Gbps (n/ac)	250 kbps	1 Mbps (v 1.2) 24 Mbps (v. 4)
# messages per day	140 12-byte	Defined by user	Defined by user	Unlimited	10 byte (W) Up to 20 byte (N) 10 byte (P)	Flexible (6 byte to 10 kbyte)	Unlimited	Unlimited	Unlimited

(Continued)

TABLE 9.5 (Continued)
Comparison of LPWAN and Other Communication Technologies

Protocol	Sigfox	LoRaWAN	NB-IoT	LTE-M	Weightless W/N/P	Ingenu	Wi-Fi	Zigbee	Bluetooth/BLE
Topology	Star	Star of star	Star	Star	Star	Star, tree	Point to hub	Star, cluster tree, mesh	Star–bus network
Battery life	8–10 years	8–10 years	7–8 years	1–2 years	<10 years	+20 years	7 days – Up to 1 year (AA battery)	100–1000+ days	Years on coin cell battery
Power efficiency	Very high	Very high	Very high	Medium	Very high	Very high	Medium	Very high	Very high
Range	10 km urban 50 km rural	2–5 km urban 15 km suburban 45 km rural	1.5 km urban 20–40 m rural	35 km 2G 200 km 3G 200 km 4G	5 km (W) 3 km (N) 2 km (P)	15 km urban	Up to 250 m	Up to 100 m	80–100 m
Scalability	Yes	Yes	Undetermined	Yes	Yes	Yes	Limited	Yes	Limited
Latency	1–30 s	1–2 s	1.4–10 s	10–15 ms	Low	>20 s	<50 ms (95th percentile)	15 ms	3 ms (BLE)
Cost	Medium	Low	High	High	Low	Medium	Low	Low	Low

that utilized an NB-IoT module to provide SMS and data transmission services over a wide range with low power consumption.

As it can be observed, the applications of LPWAN protocol are limited. This is because the standards have not yet been adopted in many areas and regions. Unlike LPWAN, the WSN legacy protocols are the first choice in many smart parking solutions. However, as the population of vehicles increases, the need for a wider communication range, more reliable, and more secure mode of communication is expected. This promises LPWAN to develop more and overcome the current existing challenges such as high complexity of interoperability between different LPWAN technologies, coexisting with other WSN protocols, and lack of standard models for large-scale applications [82,87]. As Table 9.5 suggests, almost all the LPWAN communication modules have longer than 7–8 years of battery life with higher power efficiency and wider communication range in both urban and rural areas. In rural areas, the range of the communication is wider due to the absence of obstacles such as skyscrapers, which interfere the quality of data sent and/or received. Furthermore, the most common topology used in LPWAN is of the star type where all the nodes are directly connected to a central computer or server. Every node is also connected to each other indirectly in this topology. Short-range wireless networks are mostly connected in a mesh topology to extend their range. However, the development cost and the energy usage for a large number of distributed devices make it ineffective in large-scale implementations [82] such as the multilevel parking lots. This is where LPWAN technology derives to overcome the limitations of the previous generations of wireless networks. Latency or the delay of the information from sensor nodes to the central server is quite small for legacy protocols compared to Sigfox or LoRaWAN. Nevertheless, this does not necessarily mean that they are not effective in smart parking systems. In large-scale smart parking applications where low latency is necessary, NB-IoT and LTE-M are among the best options.

9.4.2.2 Errors in Data Collection and System Reliability

Typical detection sensors in parking lots may inaccurately report the wrong number of vehicles or fail to detect it at all [88]. One of these inaccuracies is the "double counting," which occurs when the vehicle is parked in between the designated parking spots. This causes the detection of two/double vehicles instead of one. This behavior can also lead the detection sensor not to report the vehicle at all in the spot. This is mainly observed in locations where the range of the sensor is limited. Therefore, miscounting can also be a potential error/inaccuracy issue. Another type of errors is the "occlusion," where big vehicles can block the line of sight of detection sensors. And thus, smaller vehicles next to it will be hidden, and the parking lot will be considered empty. This is mostly an issue in the parking lots where two or more vehicles are entering at the same time. Another error type is the "phantom detection" error, where sensors cannot detect the correct position of the vehicle because of bouncing-off-walls signals. Other than these physically constrained errors, there might be cases where the data collected from events/observations do not conform to a predetermined pattern. In fact, data collection is useful only when we are dealing with parking prediction and/or pattern analysis in order to provide better management system and enhanced user experience. However, there is a caveat that not all sensors can behave according to a predetermined pattern. This brings the issue of data anomaly, or as the author in Ref. [89] defines it "outliers." Outliers, in general, can be caused by node malfunctioning that requires inspection or mainte-nance. In order to identify these misconducts, the authors looked at a dataset pro-vided by Worldsensing. They discovered that by looking at the dataset as a whole, the chance of missing outliers is high. They found that there exist similar data points in the dataset that share similar characteristics. And by clustering these data, they were able to easily identify the outliers while applying certain sophisticated algorithms.

Moreover, majority of the presented smart parking solutions are unable to perform system reliability checks. The system reliability is usually related to software, hard-ware, and other elements that make up the whole system to assure satisfactory per-formance in any given condition. In Ref. [90], the authors proposed a sensor-based smart parking solution with a system reliability check. They performed certain system checks to detect errors that might be caused by hardware failures or drivers' behav-iors. They created false-positive and false-negative criteria for different time inter-vals in order to assure the system reliability. In another example [91], a framework was introduced to comprise of a reputation mechanism in a mobile application-based SPS. Each time a vacant spot is reported as "empty" by the user, it receives reputa-tion score that reflects the truthfulness of the information. By doing so, the authors assure that the collected data is reliable and can be used in determining the parking vacancy. A similar application described in Ref. [92], UW-ParkAssist, integrated the collected data from the phone sensors and parking officials (or police) to avoid data manipulation by the end users. And thus, officials can manually override the status of the parking spot in the hidden settings of the application database in case of false information injection by drivers. With this verification method, the SPS is able to provide improved data quality, gathered from the users' inputs.

9.4.3 Interoperability and Data Exchange

For seamless integration of different services and technologies, the ubiquitous IoT needs to be open and able to exchange data with other platforms. In today's IoT ecosystem, we are mostly dealing with closed and vertically oriented heterogeneous systems with no vibrant collaboration [93]. Assuming connected smart parking sys-tems that can share data with other smart-city platforms in a single multi-service network is a desired challenge. For example, by analyzing the gathered informa-tion, drivers can predetermine the best route and time to reach the parking lot in the best and most efficient way. One global service, therefore, can be used on top of the already existing platforms in different regions in order to allow the same opera-tion to be performed. In order to reach such interoperability, the used data structure and interface format (i.e., syntactic interoperability), in addition to ontologies and relationships between the exchanged data (i.e., semantic interoperability), need to be carefully set. Not only data exchange between platforms but also a cross-application domain access is essential. Such ecosystem can allow the confirmation of park-ing reservation and parking finder services in an utter smart parking service [93]. Realization of interoperability in smart parking with open standards such as Open

Groups Open Messaging Interface[6] (O-MI) and Open Data Format[7] (O-DF) has been widely used for interoperable smart parking solutions. In Ref. [94], the authors offered an overview of EU's bIoTope Horizon 2020 project for economically viable IoT platforms with their proof of concept for the smart parking lot management in the coming FIFA World Cup 2022. Another proof of concept under the same EU program for EV charging with smart parking was introduced in Ref. [95]. In the following, we discuss the smart parking interoperability from the perspective of security and privacy issues.

9.4.3.1   Interoperability with Privacy and Security Aspect

Having a unique software platform and a communication backbone in the interconnected IoT era seems to be a little farfetched. And trying to make things uniform is rather a speculation. Efforts have to be spent towards embracing and managing the fast-paced IoT advancement such that there are minimal effects on the existing infrastructures and more utilization out of it. Making IoT an open world in the current heterogeneous and fragmented market is not impossible, but there are certainly some challenges that need to be addressed first. Such challenges can be (1) the lack of resources, (2) the exchanged data complexity, (3) the system proprietary, (4) and the most import one security and privacy. There are already some constraints in the existing closed and vertical IoT solutions as discussed in the previous section. Making them interoperable and able to exchange information on both lower and higher levels is admittedly a hurdle. The breach of information could pose great dangers to both users and the companies, which are responsible for these smart solutions. The capabilities of traditional internet securities embedded in the WSNs or LPWANs may not provide full protection against cyber-attacks. Trying to unify one system while maintaining the low-cost approach in IoT solutions with the current capabilities, especially in terms of the computational power, is another security concern. Interoperability of devices also brings the question of total anonymity, which is still one of the implications in developing IoT services. Many innovative IoT-related solutions depend on tracking and profiling users' movements and activities [96]. Any vulnerability could potentially lead to personal information leakage and increment in the exposure of users to other attacks, which ultimately threatens the privacy of the system users.

9.5   SOLUTIONS IN PRACTICE

In this section, we investigate various implementations of the smart parking system. Other aspects of the smart parking system in smart city applications such as the parking lot management for EV charging stations, and huge city-events which are more complex and broader in size and density have been also discussed and analyzed.

9.5.1   IMPLEMENTATION OF SMART PARKING SYSTEMS

Various implementations of the smart parking system in the literature based on their vehicular sensors as well as their mode of communication and other relevant criteria are discussed in this subsection and classified into the single-vehicle detection system and the large-scale ecosystem.

9.5.1.1 Single-Vehicle Detection Sensors

A custom infrared sensor was used in Ref. [97] to detect vehicles entering the parking spot where the results were sent to the router using Arduino by the mean of a wired connection (e.g., Ethernet). Drivers are automatically connected to the parking network and to check for parking spaces; an android mobile application based on JSON was used to determine the vacancy. This system was also used to guide drivers when they are leaving the parking spot, in addition to enabling the smart payments. On the other hand, in Ref. [98] authors used Raspberry Pi as the central server to send information to the cloud-based mobile application. In these two examples, the scalability of these systems for multilevel parking lots is of concern where the system reliability and effectiveness are major points to be considered. An example of a reliable implementation is presented in Ref. [99] where the system is comprised of both passive infrared and magnetic sensor to detect vehicles. The collected information was sampled in Java using TinyOS-2.x, and the results were reported to drivers in a web-based application. A smart parking system for commercial stretch in cities using passive infrared sensors and image detectors was discussed in Ref. [100]. It also introduced smart payment and reservation system in addition to guidance using GPS in mobile applications.

Ultrasonic is the most used sensor to detect vehicles in many areas. As presented in Ref. [101], the smart parking solution was achieved using ultrasonic vehicles' detection and Zigbee-based communication between the detection sensors and a RabbitCore microcontroller. This SPS uses the shortest-path algorithm to find the nearest parking spot and exit location near to the current vehicle location while entering/leaving the parking lot. This SPS requires the drivers to follow the directions given by the system and any deviation results in the failure of the system. Similarly in Ref. [46], ultrasonic sensors were used for detecting vehicles in a multilevel parking lot. However, additional horizontal sensors mounted on the parking walls were used for detection of the improper parking. An alarm would be triggered in order to notify the driver about this improper parking position. In this way, using three sensors per parking spot would not be feasible, and further studies should be conducted for cost-effectiveness. Ultrasonic was also used in mobile systems. The ParkNet vehicles in Ref. [102] also comprised GPS receivers and mounted ultrasonic sensors on the passenger side doors of the vehicle to detect the parking spot in urban areas. This can create a real-time map of the parking information along the street. Meanwhile, ILDs have been mostly used in traffic surveillance applications [44]. In Ref. [103], their usage in the vehicle detection was tested, and 80% success rate in detecting the magnetic signature was observed. IDLs are mostly used in conjunction with other sensors to increase the reliability and to provide traffic parameters such as the average speed, volume, and gap, which can be useful in analyzing the performance of the existing parking lots.

Magnetometer sensors are also widely used in parking lots due to their accuracy and reliability in many conditions. In Ref. [104], the authors presented their smart parking system that is comprised of magnetic sensors placed on each parking spot to detect the vacancy. It shares the results through T-sensor to base stations via Zigbee modules, and the parking vacancy information is presented through VMS. This system also offers self-healing mechanism in addition to more than 5 years of battery lifetime. In another example shown in Ref. [105], the authors used the tri-axial magnetic sensor for detecting vehicles as it enters the parking lot. The produced background noise from multiple vehicles as they approach the sensor is the issue that should be tackled in this case. A social network car park occupancy information was implemented in Ref. [106]. The system works by detecting vehicles via magnetic sensors placed at the entrance and exit of a parking space in order to inform drivers about the available car parks via LED indicators placed in several locations and connected thorough Twitter accounts and RS245&485 with the main server. A three-axis Anisotropic Magneto-Resistive (AMR) sensor for occupancy detection in parking lot for a full-fledged detection mechanism, in three phases, entering, stopping, and leaving the parking, was proposed in Ref. [107]. The authors achieved 98% detection accuracy compared to other AMR-based detection algorithms. Image processing and vehicle-license recognition techniques are used in combination with several smart parking implementations. In Ref. [108], video cameras are used to detect vehicles as they enter the parking which are sequentially checked with the help of Prewitt edge detection technique. The combined use of image processing and license recognition algorithm as previously discussed was studied in Ref. [109]. The system functions such that the characteristics of vehicles (e.g., color and license plates) are recorded on a database. Upon searching for parked vehicles by inputting the license plates, drivers can easily locate their vehicles. Since full recognition of a vehicle in some cases is not possible, classification probabilities based on the similarities of plates and colors are used for retrieval of vehicles. In Ref. [36], the authors proposed image processing in conjunction with RFID for retrieval purposes. They used RabbitCore microcontroller and Zigbee module and provided A* shortest path to assign vacant spot to drivers where they can be informed of the location of the spot via VMS or the printed map on the ticket. An adaptive and self-organized algorithm for used WSNs and RFID in locating, reserving, and charging the parking space while maintaining security measures against theft in place was provided in Ref. [10]. However, it necessitates a dedicated infrastructure in practice, which can be the most expensive in implementation.

9.5.1.2 Smart Parking in Smart City Ecosystem

Existing smart parking attempts in the smart city paradigm are either utilizing a single detection sensor, or they fall in the large-scale implementations. In large-scale scenarios, multiple technologies must orchestrate together in exchanging data to achieve a smooth interoperability between the parking lots in an IoT-based ecosystem. Smart parking in the smart city paradigm can be defined as a system that can facilitate interoperability among its subservices to provide quality of life (QoL) for urban citizens [110]. An example of a complex and large parking system can be the huge city-event parking scenario. Large events, such as sporting events, require many services to work together in order to provide an effective smart parking solution, for instance, the initiative to create an open IoT smart parking ecosystem for FIFA World Cup 2022 (in Qatar). In Ref. [111], the authors created a proof of concept while considering open standards for facilitating data interoperability and data exchange, namely, O-MI and O-DF as described in the previous section. Their concept of open IoT is applicable at stadium level as well as the city level. Another concept of a city-wide parking lot management is described in Ref. [112]. By utilizing a novel distributed algorithm and cloud computing, the authors provided a smart parking system that is context-aware and can gather information from the city or from the citizens for accurately determining the status and/or predicting the parking vacancy.

Towards cleaner environment and more energy savings, electric vehicles (EVs) are now favored worldwide. However, the lack of charging stations and/or their locations have led to severe drivers' inconvenience. Although in the future more homes and offices will be equipped with charging services, the need for a smart system to manage the charging of EVs considering time, location, and other criteria should be premeditated. Scheduling such charging stations in the smart parking context that can deliver convenience to the users and at the same time be profitable to parking owners should be carefully analyzed. In Ref. [113], the authors delivered such a scheduled system than can adapt to the traffic pattern and maximize the total utility for the owner by adjusting the time of use (TOU) pricing strategy while providing the best QoS. In order to integrate and expand the EV charging stations to the smart city ecosystem for full interoperability, like the example stated earlier, careful consideration of using the open standards must be made. To address such problems, in Ref. [95], the authors presented a case study of EV charging station via a mobile application to search and reserve these charging stations in a smart parking system. In the aforementioned examples, seamless integration of the ubiquitous IoT services and technologies must be achieved in order to realize an effective system in the ecosystem. A bi-level optimization of a smart distribution company for the parking lot owners was examined in Ref. [114]. The optimal scheduling of EV charging stations for the best of the company and the owners was evaluated by stochastic programming due to uncertainty in the parking lots. It shows the effectiveness of bi-level (distributed) approaches in comparison with centralized models. Another parking lot management system for EV was introduced in Ref. [115]. Their 24-hour trace-based model can track the mobility/patterns of EVs for the maximum revenue and EV presence in charging stations. Problems such as overstaying in charging stations or unexpected behaviors of the battery in charging can effectively disrupt and interfere with scheduling solutions. Therefore, more robust and flexible system should be researched.

9.5.2 New Applications in Smart Parking System

In this section, various emerging techniques for the smart parking system have been highlighted and discussed for a more comprehensive study. These techniques can be categorized into VANET and UAV-based techniques.

9.5.2.1 Vehicle to Everything (V2X)

With the advancement of technology and wireless communication improvements over the years, there has been a new trend in communication technology between vehicles and infrastructures known as VANET, which is a subgroup of mobile ad-hoc network (MANET) that uses vehicles as mobile nodes [116]. They are classified into three different types: vehicle-to-vehicle (v2v), vehicle-to-infrastructure (v2i), and lastly vehicle-to-roadside (V2R). The mac layer standard of VANET is in the IEEE 802.11p standard with a connectivity range of 100–1000 m and 27.0 Mbps data rate that promises fast and reliable method of communication. However, they are more expensive than other protocols [117]. Onboard Unit (OBU) installed in vehicles by manufactures and the Roadside Unit (RSU) installed near the road providing both road safety and a mode of communication between the vehicle and the roadside infrastructure are the requirements in this type of SPS [4]. VANET is also used in smart parking systems where vehicles can detect parking occupancy and report the location to other vehicles or to roadside units in a two-way communication manner [44]. As presented in Ref. [118], DIG-Park, with the combination of V2I and RSU along with distance geometry algorithms to find the nearest parking spot can be used for both indoor and outdoor applications. The authors provided a real-time parking navigation as well as an anti-theft and anti-collision protection that make use of OBU and RSU. Similar work was carried out in Ref. [4]. However, this model represented a large parking lot, which adds more complexity compared to the previous work. SDN was also overviewed and examined in Ref. [119] as to provide more flexible, programmable network in VANET paradigm that can overcome some of the difficulties and complexities in the existing vehicular environment. VANET possibilities are enormous. However, some problems and difficulties arise when using this system such as the problems in frequency and bandwidth spectrum [120] because VANET is operated in the 75 MHz band at 5.9 GHz where in some countries this band is used for military and radar systems. Another issue lies within the routing protocol where multi-hop communication between vehicles makes the current routing protocols incompatible with VANET such as the current protocol in MANET. Therefore, the current research is mostly focused on creating a new VANET-based routing protocol.

9.5.2.2 Unmanned Aerial Vehicles (UAVs)

UAVs, also known as drones, are also becoming a trend in today's IoT applications. They are mainly used in wireless communication and serve as a backbone for the network connectivity in places where communication range is limited or it fails to work in case of a disaster. Equipping drones with the vehicle detection sensors as discussed previously not only can provide the benefit of having a one-package smart parking system with all the advantages of a regular smart parking system, but also can overcome the shortcomings of the sensors in certain areas. Larger detection range, ability to carry multiple sensors, high precision, flexibility in deployment, high mobility, and many others are just a few advantages showing the ascendency of drones compared to static vehicle detection sensors. A UAV-assisted smart parking example was studied in Ref. [121]. In this work, the authors utilized generative adversarial network (GAN) for detection and prediction of the parking vacancy in a vision-based approach. High precision of their UAVs shows the superiority of their detection algorithm and obstacle avoidance technique. Another similar solution was proposed in Ref. [122]. The authors in their study used the markers printed on the asphalt as their parameter to check for parking vacancy from their single drone application. Low light, harsh weather conditions, and collision avoidance between a team of drones are some of the problems that need to be examined while implementing a drone-based smart parking solution. The coextensive UAV-assisted VANET in the future of smart parking in IoT era can be one of the approaches that offers a synergistic combination of the two described technologies.

9.5.3   HYBRID SYSTEM

A hybrid model can exist when more than one type of sensors are used for detecting vehicles, or when both wired and wireless communication modes are used in conjunction. Integrating multiple data source into one provides more reliable and faster way for detecting vehicles where in case of failure of one sensor node, the integrity of the system would not collapse. Streetline application [123] in hybrid smart parking is the perfect example of such a system. In Ref. [123], authors incorporated magnetometer, light sensors, CCTV, and mobile phone applications to detect and report the vacancy of parking spots in a large parking lot. They also offer a dynamic pricing strategy for their payment systems. In another implementation of the hybrid smart parking in Ref. [124], the authors presented microaerial vehicle (MAV) indoor application for smart parking system which is equipped with two ultrasonic sensors and two cameras for the parking spots detection. NFC tag and QR codes are also used for locating the MAV, for recording and updating the latest status of the parking spot and as a way for drivers to record their parking information. In case of hybrid wired and wireless communication, the authors in Ref. [16] presented a hybrid system that uses presence sensors and RFID tags for billing purposes which are both provided with wired and wireless capabilities, using 802.15.4/Zigbee standards.

In general, as summarized in Table 9.6, the hybrid system provides a flexible solution to smart parking applications where the functionality of the system may change in the future. Furthermore, the scalability of such systems in large applications, such as the urban parking, is high. Deficiencies of a single-sensor approach in this scenario can be overcome. Hybrid systems can be the future of smart parking solutions, and the worldwide usage of these systems can be anticipated in the future prospect

---

TABLE 9.6

Comparison of the Hybrid and Single-Sensor Smart Parking Systems

	Parking System Model	
Characteristics	**Hybrid [10,68,70,71,90]**	**Single Sensor [3,21,55–60,88–92]**
Large-scale application	✓	Limited
Flexibility	✓	Limited
Cost effectiveness	✓	Sensor dependent
General reliability	✓	Very low
Service providers	✓	Limited
Overall recommendation	Medium/large scale	Small scale

✓ Indicates only applicability in certain cases (applications) as discussed in each subsection.

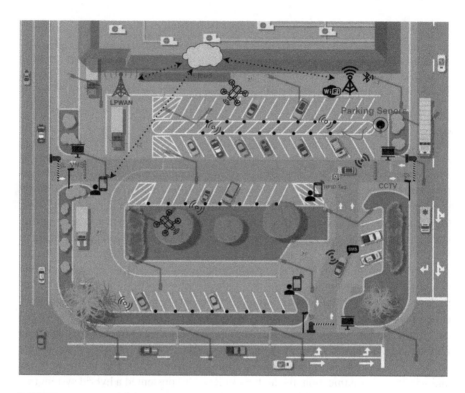

FIGURE 9.3 A model for a hybrid smart parking system.

of the IoT paradigms. The general concept of a hybrid smart parking application is shown in Figure 9.3. In this model, a series of parking sensors are installed throughout the parking lot. We also recommend the usage of UAVs to further enhance the detection and navigation. All systems are connected and exchanging information using open standards with 5G or LPWAN communication modules. The gates are controlled by CCTV, which can read the license plate of the vehicles. This can be used to estimate the total parking time and using algorithms to automatically determine the fee. The fee is also dynamically adjusted based on the traffic, and the time of day the parking is utilized. Users have options to pay the fee in various ways. Many VMSs are also installed to indicate useful information to the drivers. By introducing a cloud-based smart parking system, we can assure a seamless integration with many services such that users can reserve and find the best route to reach the parking lot with a guaranteed parking space at any time of the day. By incorporating open APIs and open standards with the power of open-sourced software and services, we can introduce a fast and safe smart parking system that can easily integrate with the other elements in the smart city ecosystem.

Although there are challenges in these systems, such as the requirement to handle multiple data sources entailing complex algorithms and the requirement of having a safe and fast method of communication between multiple sensors to decrease the latency of data, at the end, the outcome of these hybrid systems is beneficial to both the user and the parking lot owners. Because these systems are the most effective in providing useful services such as integrated smart parking reservation, smart payments, automatic gate control, and dynamic pricing strategies based on the availability of the parking spot and other factors that maximize utilization of the parking lots during the peak and off-peak hours.

The cost effectiveness of the hybrid approach is generally substantiated by the type and the number of sensors used. Single sensor smart parking solutions as described earlier in this chapter may vary in terms of the accuracy of the collected data. Moreover, in these applications, in case of failure of the sensor, the entire application may fail which results in discomfort and economical cost. This is where hybrid solutions can come into the play. Although the initial capital cost may deter the use of this approach, in the long term, the benefits of using such a system can easily be realized. In fact, the main components of the hybrid system can easily integrate with the emerging smart city components. The large-scale deployment of this hybrid model paves the way towards more robust and flexible solutions, which can provide innovative services in cost-effective manners.

9.6   OPEN RESEARCH ISSUES

In this chapter, a complete survey of existing smart parking solutions was discussed where several types of sensors in terms of their functionalities and their strengths and weaknesses were summarized. As discussed in this chapter, we cannot find a single best sensor that is applicable for all smart parking solutions. Some may fail in extreme weather, or they required to be placed inside pavements, which have its complications. Others may have limited use in certain applications due to privacy and security issues. Therefore, the efforts to diminish these shortcomings in the information collection are now put into mobile sensing devices where some of the limits or barriers of fixed sensors can be addressed. However, due to irregularities in spatiotemporal coverage and the issue of big data that are being transmitted, aggregation of such data and predicting the availability of the parking spots are the challenges that if tackled correctly could be the new future of the mobile sensing devices.

Moreover, having a reliable, fast, and secure mode of communication in smart parking system is another problem. As discussed in this chapter, the new era of LPWAN communication protocol seems to the future of commutations as they provide large area coverage, low-power consumption, and high battery lifetime, as well as higher security measures compared to the legacy communication modules. Large-scale applications of LPWAN are still under study as there is still the issue of interoperability and coexisting with other communication technologies used in WSNs. A large amount of data and big packet sizes that are being transmitted require a steadfast, end-to-end encrypted communication. As it was estimated that by 2023, there would be 20–40 billion connected devices. In this scenario, there could be a bottleneck in the existing communications infrastructures [125].

The fifth-generation (5G) mobile network is expected to be the center of the emerging IoT devices in the near future. With the ever-increasing applications in cloud computing and smart devices, 5G can promise to address some, if not all, the current issues of telecommunications. Studies are now underway to integrate existing devices with 5G wireless communication.

Connected vehicles are also another interesting option that is being used in several applications of smart parking solutions. Mobile sensors connected to vehicles and smart mobile applications that could identify parking spots are now introduced by car manufactures. The idea of connecting everything including our cars to each other may sound overwhelming as it can provide attractive services such as real-time navigation and crowdsourced information. However, the current technologies in connected vehicles may limit the full potential of these types of applications particularly in urban areas [126]. Varying speed, specific bandwidth for communication, and the need for a better routing protocol and a high-speed communication technology to send and receive information and in general the QoS including delays are the challenges in the current era of connected vehicles. In Ref. [106], the authors proposed a mathematical model for a new-generation of forwarding QoS approaches that enables the allocation of the optimal path that satisfies the QoS parameters while supporting a wide range of communication-intensive IoT applications. The model is used to investigate the effects of multi-hop communication on a traffic system model designed with a Markov discrete-time M/M/1 queuing model, applicable to green deployment of duty-cycle sensor nodes. The authors, in this paper, have presented analytical formulation for the bit error-rate, and a critical path-loss model is defined to the specified level of trust among the most frequently used nodes. Additionally, the authors have addressed the degree of irregularity parameter for promoting adaptation to geographic switching with respect to two categories of transmission in distributed systems: hop-by-hop and end-to-end retransmission schemes.

Nowadays, most of the deployed smart parking systems rely on battery-powered sensors and wireless communication modules. Although there are possibilities of implementing energy aware algorithms or routing protocols that could effectively reduce the overall energy consumption as discussed in Ref. [127], at some point due to size reduction of circuit boards and increase in big data we need to look at the potentials of energy harvesting modules. These modules use power generating elements such as solar cells, piezoelectric elements, and thermoelectric elements to generate electricity by converting different energy sources such as light, vibration, and heat, respectively. ParkHere [128] is a self-powered parking sensor that uses the weight of vehicles to power a microgenerator which at the same time can send its information of the occupancy of the parking spot to the server via mobile radio.

As more and more things connect to each other, it is imperative to IoT ecosystem for all systems in general to be able communicate and exchange information to one another. Developing a multifaceted technology to improve the existing infrastructures while providing maximum security and privacy measures with total and fully controlled management mechanism and with interoperability functionalities is vital for the future advancement of the ecosystem. Nowadays, extensive time and efforts are spent and devoted to AI, machine learning (ML), and interoperability in order to provide solutions to the aforementioned challenges in SPS. Many aspects of the ecosystem need to work coherently to be able to overcome such difficulties. Such solutions would not only provide a better QoS to users of the system but also assure the efficiency and the integrity of all the elements that represent the system.

9.7 CONCLUDING REMARKS

As the population of urban area increases leading to traffic congestions and other problems, the need for parking spots is inevitable. In the age of IoT and smart city ecosystems, it is not difficult to believe why a smart and innovative solution is considered to pave way towards a more sustainable future for cities. Hence, to improve the current parking system and to address significant issues in over-crowded cities, the smart parking system has been overviewed. Many parameters must be well studied and analyzed before implementing any smart parking system. Therefore, a comprehensive survey of the current state of smart parking systems including classification of the parking system, major vehicle detection technologies, and communication module have been presented in this work. The objective of this survey was to offer an insight into new research attempts in the intelligent transportation system. We looked at different elements of the smart parking system and explained thoroughly the hardware and software aspects of this application. Software aspects of the smart parking were presented, and several features such as parking prediction, path optimization, and parking assignment, and how the collected information can enhance the experience of parking operators and drivers were introduced. Different tables were generated in order to compare several key factors about the main elements in a smart parking system, including sensors and communication modules. Moreover, an overview of the associated data security/privacy, and new trends in interoperability and data exchange have been discussed. Next, emerging technologies in smart parking systems, namely, the V2X and UAVs, have been presented. The concept of cloud-based hybrid models has been suggested in order to solve key issues in smart parking applications. As a future work, we plan to further investigate our cloud-based hybrid concept with the help of interoperability and drones in order to provide a proof of concept for the next generation of the smart parking systems.

NOTES

1 http://www.libelium.com/products/plug-sense/.
2 http://www.huawei.com/minisite/iot/en/smart-parking.html.
3 http://www.zte.com.cn/global/.
4 https://www.nwave.io/parking-technology/.
5 https://www.telensa.com/smart-parking/.
6 http://www.opengroup.org/iot/omi/.
7 http://www.opengroup.org/iot/odf/.

REFERENCES

1. G. Cookson, "Parking pain – INRIX offers a silver bullet," *INRIX – INRIX* [Online]. Available: http://inrix.com/blog/2017/07/parkingsurvey/. [Accessed: 21-Nov-2017].
2. M. Gohar, M. Muzammal, and A. Ur Rahman, "SMART TSS: Defining transportation system behavior using big data analytics in smart cities," *Sustain. Cities Soc.*, vol. 41, pp. 114–119, Aug. 2018 [Online]. Available: https://linkinghub.elsevier.com/retrieve/pii/S2210670717309757. [Accessed: 24-Feb-2019].

3. S. E. Bibri, "The IoT for smart sustainable cities of the future: An analytical framework for sensor-based big data applications for environmental sustainability," *Sustain. Cities Soc.*, vol. 38, pp. 230–253, Apr. 2018 [Online]. Available: https://linkinghub.elsevier.com/retrieve/pii/S2210670717313677. [Accessed: 25-Feb-2019].

4. R. Lu, X. Lin, H. Zhu, and X. Shen, "SPARK: A new VANET-based smart parking scheme for large parking lots," in *IEEE INFOCOM 2009*, 2009, pp. 1413–1421.

5. D. Uckelmann, M. Harrison, and F. Michahelles, "An architectural approach towards the future Internet of Things," in *Architecting the Internet of Things*, D. Uckelmann, M. Harrison, and F. Michahelles, Eds. Berlin, Heidelberg: Springer, 2011, pp. 1–24 [Online]. Available: doi: 10.1007/978-3-642-19157-2_1.

6. F. Al-Turjman, "Intelligence and Security in Big 5G-oriented IoNT: An Overview," *Elsevier Future Gener. Comp. Sy.*, vol. 102, no. 1, pp. 357–368, 2020.

7. M. Chandrahasan, A. Mahadik, T. Lotlikar, M. Oke, and A. Yeole, "Survey on different smart parking techniques," *Int. J. Comput. Appl.*, vol. 137, no. 13, pp. 17–21, Mar. 2016 [Online]. Available: http://www.ijcaonline.org/research/volume137/number13/chandrahasan-2016-ijca-908920.pdf. [Accessed: 22-Feb-2019]

8. G. Revathi and V. R. S. Dhulipala, "Smart parking systems and sensors: A survey," in *2012 International Conference on Computing, Communication and Applications*, 2012, pp. 1–5.

9. M. Y. I. Idris, Y. Y. Leng, E. M. Tamil, N. M. Noor, and Z. Razak, "Car park system: A review of smart parking system and its technology," *Inf. Technol. J.*, vol. 8, no. 2, pp. 101–113, Feb. 2009 [Online]. Available: http://www.scialert.net/abstract/?doi=itj.2009.101.113. [Accessed: 07-Nov-2017].

10. A. Hilmani, A. Maizate, and L. Hassouni, "Designing and managing a smart parking system using wireless sensor networks," *J. Sens. Actuat. Netw.*, vol. 7, no. 2, p. 24, Jun. 2018 [Online]. Available: http://www.mdpi.com/2224-2708/7/2/24. [Accessed: 21-Feb-2019].

11. S. Faheem, S. A. Mahmud, G. M. Khan, M. Rahman, and H. Zafar, "A survey of intelligent car parking system," *J. Appl. Res. Technol.*, vol. 11, no. 5, pp. 714–726, Oct. 2013 [Online]. Available: http://www.sciencedirect.com/science/article/pii/S1665642313715803. [Accessed: 10-Dec-2017].

12. K. Hassoune, W. Dachry, F. Moutaouakkil, and H. Medromi, "Smart parking systems: A survey," in *2016 11th International Conference on Intelligent Systems: Theories and Applications (SITA)*, Mohammedia, Morocco, 2016, pp. 1–6 [Online]. Available: http://ieeexplore.ieee.org/document/7772297/. [Accessed: 22-Feb-2019].

13. T. Lin, H. Rivano, and F. Le Mouel, "A survey of smart parking solutions," *IEEE Trans. Intell. Transp. Syst.*, vol. 18, no. 12, pp. 3229–3253, Dec. 2017 [Online]. Available: http://ieeexplore.ieee.org/document/7895130/. [Accessed: 22-Feb-2019].

14. E. Polycarpou, L. Lambrinos, and E. Protopapadakis, "Smart parking solutions for urban areas," in *2013 IEEE 14th International Symposium on "A World of Wireless, Mobile and Multimedia Networks" (WoWMoM)*, 2013, pp. 1–6.

15. J. Chinrungrueng, U. Sunantachaikul, and S. Triamlumlerd, "Smart parking: An application of optical wireless sensor network," in *2007 International Symposium on Applications and the Internet Workshops*, 2007, p. 66.

16. A. Bagula, L. Castelli, and M. Zennaro, "On the design of smart parking networks in the smart cities: An optimal sensor placement model," *Sensors*, vol. 15, no. 7, pp. 15443–15467, Jun. 2015 [Online]. Available: http://www.mdpi.com/1424-8220/15/7/15443. [Accessed: 28-Feb-2018].

17. Z. Hui-ling, X. Jian-min, T. Yu, H. Yu-cong, and S. Ji-feng, "The research of parking guidance and information system based on dedicated short range communication," in *Proceedings of the 2003 IEEE International Conference on Intelligent Transportation Systems*, 2003, vol. 2, pp. 1183–1186.

18. Y. Qian and G. Hongyan, "Study on parking guidance and information system based on intelligent mobile phone terminal," in *2015 8th International Conference on Intelligent Computation Technology and Automation (ICICTA)*, 2015, pp. 871–874.

19. M. Buntić, E. Ivanjko, and H. Gold, "ITS supported parking lot management," in *International Conference on Traffic and Transport Engineering-Belgrade*. Belgrade, Serbia, 2012.

20. Y.-C. Shiue, J. Lin, and S.-C. Chen, "A study of geographic information system combining with GPS and 3G for parking guidance and information system," *City*, vol. 65, no. 6, p. 9, 2010.

21. M. Chen and T. Chang, "A parking guidance and information system based on wireless sensor network," in *2011 IEEE International Conference on Information and Automation*, 2011, pp. 601–605.

22. M. Patil and V. N. Bhonge, "Parking guidance and information system using RFID and Zigbee," *Int. J. Eng. Res. Technol.*, vol. 2, no. 4, Apr. 2013. ISSN 2278-0181.

23. E. Kokolaki, M. Karaliopoulos, and I. Stavrakakis, "Opportunistically assisted parking service discovery: Now it helps, now it does not," *Pervasive Mob. Comput.*, vol. 8, no. 2, pp. 210–227, Apr. 2012 [Online]. Available: http://www.sciencedirect.com/science/article/pii/S1574119211000782. [Accessed: 10-Dec-2017].

24. M. Ş. Kuran, A. C. Viana, L. Iannone, D. Kofman, G. Mermoud, and J. P. Vasseur, "A smart parking lot management system for scheduling the recharging of electric vehicles," *IEEE Trans. Smart Grid*, vol. 6, no. 6, pp. 2942–2953, Nov. 2015.

25. K. Raichura and N. Padhariya, "edPAS: Event-Based dynamic parking allocation system in vehicular networks," in *2014 IEEE 15th International Conference on Mobile Data Management*, 2014, vol. 2, pp. 79–84.

26. D. Bong, K. C. Ting, and K. C. Lai, "Integrated approach in the design of car park occupancy information system (COINS)," *IAENG Int. J. Comput. Sci.*, vol. 35, pp. 7–14, Jan. 2008.

27. M. Buntić, E. Ivanjko, and H. Gold, "ITS supported parking lot management," presented at the *International Conference on Traffic and Transport Engineering*, Belgrade, 2012.

28. F. Al-Turjman, and C. Altrjman, "Enhanced Medium Access for Traffic Management in Smart-cities' Vehicular-Cloud", *IEEE Intelligent Transportation Systems Magazine*, 2020. DOI: 10.1109/MITS.2019.2962144.

29. I. Benenson, K. Martens, and S. Birfir, "PARKAGENT: An agent-based model of parking in the city," *Comput. Environ. Urban Syst.*, vol. 32, no. 6, pp. 431–439, Nov. 2008 [Online]. Available: http://www.sciencedirect.com/science/article/pii/S0198971508000689. [Accessed: 10-Dec-2017].

30. S.-Y. Chou, S.-W. Lin, and C.-C. Li, "Dynamic parking negotiation and guidance using an agent-based platform," *Expert Syst. Appl.*, vol. 35, no. 3, pp. 805–817, Oct. 2008 [Online]. Available: http://www.sciencedirect.com/science/article/pii/S095741740700293X. [Accessed: 10-Dec-2017].

31. C. J. Rodier and S. A. Shaheen, "Transit-based smart parking: An evaluation of the San Francisco Bay area field test," *Transp. Res. Part C Emerg. Technol.*, vol. 18, no. 2, pp. 225–233, Apr. 2010 [Online]. Available: http://www.sciencedirect.com/science/article/pii/S0968090X09001120. [Accessed: 09-Dec-2017].

32. S. Pal and V. Singh, "GIS based transit information system for metropolitan cities in India," in *The Proceedings of Geospatial World Forum*, 2011, pp. 18–21.

33. Z.-R. Peng, "A methodology for design of a GIS-based automatic transit traveler information system," *Comput. Environ. Urban Syst.*, vol. 21, no. 5, pp. 359–372, Sep. 1997 [Online]. Available: http://www.sciencedirect.com/science/article/pii/S0198971598000064. [Accessed: 09-Dec-2017].

34. E. Kokolaki, G. Kollias, M. Papadaki, M. Karaliopoulos, and I. Stavrakakis, "Opportunistically-assisted parking search: A story of free riders, selfish liars and bona fide mules," in *2013 10th Annual Conference on Wireless On-demand Network Systems and Services (WONS)*, 2013, pp. 17–24.

35. D. Thierry, I. Sergio, L. Sylvain, and C. Nicolas, "Sharing with caution: Managing parking spaces in vehicular networks," *Mob. Inf. Syst.*, no. 1, pp. 69–98, 2013 [Online]. Available: http://www.medra.org/servlet/aliasResolver?alias=iospress&genre=article& issn=1574-017X&volume=9&issue=1&spage=69&doi=10.3233/MIS-2012-0149. [Accessed: 10-Dec-2017].

36. K. Malik, M. Ahmad, S. Khalid, H. Ahmad, F. Al-Turjman, and S. Jabbar, "Image and Command Hybrid Model for Vehicles Control using Internet of Vehicles (IoV)," *Wiley Trans. Emerg. Telecommun. Technol.*, vol. 31, no. 5, 2020.

37. N. V. Juliadotter, "Hacking smart parking meters," in *2016 International Conference on Internet of Things and Applications (IOTA)*, 2016, pp. 191–196.

38. K. Mouskos, M. Boile, and N. A. Parker, "Technical solutions to overcrowded park and ride facilities," New Jersey Department of Transportation, FHWA-NJ-2007-011, 2007.

39. N. H. H. M. Hanif, M. H. Badiozaman, and H. Daud, "Smart parking reservation system using short message services (SMS)," in *2010 International Conference on Intelligent and Advanced Systems*, 2010, pp. 1–5.

40. V. Menon, S. Jacob, S. Joseph, P. Sehdev, M. Khosravi, and F. Al-Turjman, "An IoT-Enabled Intelligent Automobile System for Smart Cities", *Elsevier Internet of Things*, 2020. DOI. 10.1016/j.iot.2020.100213.

41. S. Noor, R. Hasan, and A. Arora, "ParkBid: An incentive based crowdsourced bidding service for parking reservation," in *2017 IEEE International Conference on Services Computing (SCC)*, 2017, pp. 60–67.

42. P. Sauras-Perez, A. Gil, and J. Taiber, "ParkinGain: Toward a smart parking application with value-added services integration," in *2014 International Conference on Connected Vehicles and Expo (ICCVE)*, 2014, pp. 144–148.

43. Federal Highway Administration, "Traffic control systems handbook: Chapter 6 detectors – FHWA office of operations." [Online]. Available: https://ops.fhwa.dot.gov/ publications/fhwahop06006/chapter_6.htm#t62fnb. [Accessed: 04-Jan-2018].

44. T. Lin, H. Rivano, and F. L. Mouël, "A survey of smart parking solutions," *IEEE Trans. Intell. Transp. Syst.*, vol. 18, no. 12, pp. 3229–3253, Dec. 2017.

45. F. Al-Turjman, "Mobile couriers' selection for the smart-grid in smart cities' pervasive sensing," *Future Gener. Comp. Syst.*, vol. 82, no. 1, pp. 327–341, 2018.

46. A. Kianpisheh, N. Mustaffa, P. Limtrairut, and P. Keikhosrokiani, "Smart parking system (SPS) architecture using ultrasonic detector," *Int. J. Softw. Eng. Appl.*, vol. 6, no. 3, pp. 55–58, 2012.

47. A. O. Kotb, Y. C. Shen, and Y. Huang, "Smart parking guidance, monitoring and reservations: A review," *IEEE Intell. Transp. Syst. Mag.*, vol. 9, no. 2, pp. 6–16, Summer 2017.

48. P. T. Martin, Y. Feng, and X. Wang, *Detector Technology Evaluation*. Fargo, ND: Mountain-Plains Consortium, 2003.

49. B. Song, H. Choi, and H. S. Lee, "Surveillance tracking system using passive infrared motion sensors in wireless sensor network," in *2008 International Conference on Information Networking*, 2008, pp. 1–5.

50. G. M. Someswar, R. B. Dayananda, S. Anupama, J. Priyadarshini, and A. A. Shariff, "Design & development of an autonomic integrated car parking system," *Compusoft*, vol. 6, no. 3, p. 2309, 2017.

51. M. Bachani, U. M. Qureshi, and F. K. Shaikh, "Performance analysis of proximity and light sensors for smart parking," *Procedia Comput. Sci.*, vol. 83, pp. 385–392, Jan. 2016 [Online]. Available: http://www.sciencedirect.com/science/article/pii/ S1877050916302332. [Accessed: 28-Feb-2018].

52. M. Arab and T. Nadeem, "MagnoPark – Locating on-street parking spaces using magnetometer-based pedestrians' smartphones," in *2017 14th Annual IEEE International Conference on Sensing, Communication, and Networking (SECON)*, 2017, pp. 1–9.

53. Z. Pala and N. Inanc, "Smart parking applications using RFID technology," in *2007 1st Annual RFID Eurasia*, 2007, pp. 1–3.

54. E. Karbab, D. Djenouri, S. Boulkaboul, and A. Bagula, "Car park management with networked wireless sensors and active RFID," in *2015 IEEE International Conference on Electro/Information Technology (EIT)*, 2015, pp. 373–378.

55. A. Khanna and R. Anand, "IoT based smart parking system," in *2016 International Conference on Internet of Things and Applications (IOTA)*, 2016, pp. 266–270.

56. J. Rico, J. Sancho, B. Cendon, and M. Camus, "Parking easier by using context information of a smart city: Enabling fast search and management of parking resources," in *2013 27th International Conference on Advanced Information Networking and Applications Workshops*, 2013, pp. 1380–1385.

57. H. Zhu, J. Liu, L. Peng, and H. Li, "Real-time parking guidance model based on Stackelberg game," in *2017 IEEE International Conference on Information and Automation (ICIA)*, 2017, pp. 888–893.

58. W. Liang, Y. Zhang, J. Hu, and X. Wang, "A personalized route guidance approach for urban travelling and parking to a shopping mall," in *2017 4th International Conference on Transportation Information and Safety (ICTIS)*, 2017, pp. 319–324.

59. K. Hantrakul, S. Sitti, and N. Tantitharanukul, "Parking lot guidance software based on MQTT protocol," in *2017 International Conference on Digital Arts, Media and Technology (ICDAMT)*, 2017, pp. 75–78.

60. X. Zhang, L. Yu, Y. Wang, G. Xue, and Y. Xu, "Intelligent travel and parking guidance system based on Internet of vehicle," in *2017 IEEE 2nd Advanced Information Technology, Electronic and Automation Control Conference (IAEAC)*, 2017, pp. 2626–2629.

61. L. Xie, J. Liu, C. Miao, and M. Liu, "Study of method on parking guidance based on video," in *2016 IEEE 11th Conference on Industrial Electronics and Applications (ICIEA)*, 2016, pp. 1394–1399.

62. I. Aydin, M. Karakose, and E. Karakose, "A navigation and reservation based smart parking platform using genetic optimization for smart cities," in *2017 5th International Istanbul Smart Grid and Cities Congress and Fair (ICSG)*, 2017, pp. 120–124.

63. A. Houissa, D. Barth, N. Faul, and T. Mautor, "A learning algorithm to minimize the expectation time of finding a parking place in urban area," in *2017 IEEE Symposium on Computers and Communications (ISCC)*, 2017, pp. 29–34.

65. Y. Ji, D. Tang, P. Blythe, W. Guo, and W. Wang, "Short-term forecasting of available parking space using wavelet neural network model," *IET Intell. Transp. Syst.*, vol. 9, no. 2, pp. 202–209, 2015.

66. F. Caicedo, C. Blazquez, and P. Miranda, "Prediction of parking space availability in real time," *Expert Syst. Appl.*, vol. 39, no. 8, pp. 7281–7290, Jun. 2012 [Online]. Available: http://www.sciencedirect.com/science/article/pii/S0957417412001042. [Accessed: 20-Mar-2018].

67. J. S. Leu and Z. Y. Zhu, "Regression-based parking space availability prediction for the Ubike system," *IET Intell. Transp. Syst.*, vol. 9, no. 3, pp. 323–332, 2015.

64. Y. Zheng, S. Rajasegarar, and C. Leckie, "Parking availability prediction for sensor-enabled car parks in smart cities," in *2015 IEEE Tenth International Conference on Intelligent Sensors, Sensor Networks and Information Processing (ISSNIP)*, 2015, pp. 1–6.

68. M. Maric, D. Gracanin, N. Zogovic, N. Ruskic, and B. Ivanovic, "Parking search optimization in urban area," *Int. J. Simul. Model.*, vol. 16, pp. 195–206, Jun. 2017.

69. A. Moradkhany, P. Yi, I. Shatnawi, and K. Xu, "Minimizing parking search time on urban university campuses through proactive class assignment," *Transp. Res. Rec. J. Transp. Res. Board*, vol. 2537, pp. 158–166, Jan. 2015 [Online]. Available: http://trrjournalonline.trb.org/doi/abs/10.3141/2537-17. [Accessed: 21-Mar-2018].

70. M. Rybarsch, M. Aschermann, F. Bock, A. Goralzik, F. Köster, M. Ringhand, and A. Trifunović, "Cooperative parking search: Reducing travel time by information exchange among searching vehicles," in *2017 IEEE 20th International Conference on Intelligent Transportation Systems (ITSC)*, 2017, pp. 1–6.

71. S. Banerjee and H. Al-Qaheri, "An intelligent hybrid scheme for optimizing parking space: A tabu metaphor and rough set based approach," *Egypt. Inform. J.*, vol. 12, no. 1, pp. 9–17, Mar. 2011 [Online]. Available: http://www.sciencedirect.com/science/article/pii/S1110866511000077. [Accessed: 21-Mar-2018].

72. "Internet of Things outlook – Ericsson," *Ericsson.com*, 09-Nov-2017 [Online]. Available: https://www.ericsson.com/en/mobility-report/reports/november-2017/internet-of-things-outlook. [Accessed: 26-Dec-2017].

73. I. Chatzigiannakis, A. Vitaletti, and A. Pyrgelis, "A privacy-preserving smart parking system using an IoT elliptic curve based security platform," *Comput. Commun.*, vol. 89–90, no. Supplement C, pp. 165–177, Sep. 2016 [Online]. Available: http://www.sciencedirect.com/science/article/pii/S014036641630072X. [Accessed: 26-Dec-2017].

74. T. Braun, B. C. M. Fung, F. Iqbal, and B. Shah, "Security and privacy challenges in smart cities," *Sustain. Cities Soc.*, vol. 39, pp. 499–507, May 2018 [Online]. Available: https://linkinghub.elsevier.com/retrieve/pii/S2210670717310272. [Accessed: 24-Feb-2019].

75. J. Ni, K. Zhang, Y. Yu, X. Lin, and X. S. Shen, "Privacy-preserving smart parking navigation supporting efficient driving guidance retrieval," *IEEE Trans. Veh. Technol.*, vol. PP, no. 99, p. 1, 2018.

76. R. Lu, X. Lin, H. Zhu, and X. Shen, "An intelligent secure and privacy-preserving parking scheme through vehicular communications," *IEEE Trans. Veh. Technol.*, vol. 59, no. 6, pp. 2772–2785, Jul. 2010.

77. H. Wang and W. He, "A reservation-based smart parking system," in *2011 IEEE Conference on Computer Communications Workshops (INFOCOM WKSHPS)*, 2011, pp. 690–695.

78. J. de C. Silva, J. J. P. C. Rodrigues, A. M. Alberti, P. Solic, and A. L. L. Aquino, "LoRaWAN #x2014; A low power WAN protocol for Internet of Things: A review and opportunities," in *2017 2nd International Multidisciplinary Conference on Computer and Energy Science (SpliTech)*, 2017, pp. 1–6.

79. M. Collotta, G. Pau, T. Talty, and O. K. Tonguz, "Bluetooth 5: A concrete step forward towards the IoT," ArXiv171100257 Cs, Nov. 2017.

80. O. Wellnitz and L. Wolf, "On latency in IEEE 802.11-based wireless ad-hoc networks," in *IEEE 5th International Symposium on Wireless Pervasive Computing 2010*, 2010, pp. 261–266.

81. R. S. Sinha, Y. Wei, and S.-H. Hwang, "A survey on LPWA technology: LoRa and NB-IoT," *ICT Express*, vol. 3, no. 1, pp. 14–21, Mar. 2017 [Online]. Available: http://www.sciencedirect.com/science/article/pii/S2405959517300061. [Accessed: 19-Dec-2017].

82. U. Raza, P. Kulkarni, and M. Sooriyabandara, "Low power wide area networks: An overview," *IEEE Commun. Surv. Tutor.*, vol. 19, no. 2, pp. 855–873, Secondquarter 2017.

83. S. Al-Sarawi, M. Anbar, K. Alieyan, and M. Alzubaidi, "Internet of Things (IoT) communication protocols: Review," in *2017 8th International Conference on Information Technology (ICIT)*, 2017, pp. 685–690.

84. A. Asaduzzaman, K. K. Chidella, and M. F. Mridha, "A time and energy efficient parking system using Zigbee communication protocol," in *SoutheastCon 2015*, 2015, pp. 1–5.

85. M. Lauridsen, H. Nguyen, B. Vejlgaard, I. Z. Kovacs, P. Mogensen, and M. Sorensen, "Coverage comparison of GPRS, NB-IoT, LoRa, and SigFox in a 7800 km #x000B2; Area," in *2017 IEEE 85th Vehicular Technology Conference (VTC Spring)*, 2017, pp. 1–5.

86. J. Shi, L. Jin, J. Li, and Z. Fang, "A smart parking system based on NB-IoT and third-party payment platform," in *2017 17th International Symposium on Communications and Information Technologies (ISCIT)*, 2017, pp. 1–5.

87. A. Lavric and V. Popa, "Internet of Things and LoRa #x2122; Low-power wide-area networks: A survey," in *2017 International Symposium on Signals, Circuits and Systems (ISSCS)*, 2017, pp. 1–5.

88. M. Dalgleish, *Highway Traffic Monitoring and Data Quality*. Norwood, MA: Artech House, 2008.

89. N. Piovesan, L. Turi, E. Toigo, B. Martinez, and M. Rossi, "Data analytics for smart parking applications," *Sensors*, vol. 16, no. 10, Sep. 2016.

90. A. Araújo, R. Kalebe, G. Girão, I. Filho, K. Gonçalves, and B. Neto, "Reliability analysis of an IoT-based smart parking application for smart cities," in *2017 IEEE International Conference on Big Data (Big Data)*, 2017, pp. 4086–4091.

91. S. Gupte and M. Younis, "Participatory-sensing-enabled efficient parking management in modern cities," in *2015 IEEE 40th Conference on Local Computer Networks (LCN)*, 2015, pp. 241–244.

92. J. Villalobos, B. Kifle, D. Riley, and J. U. Quevedo-Torrero, "Crowdsourcing automobile parking availability sensing using mobile phones," *UWM Undergrad. Res. Symp.*, pp. 1–7, 2015.

93. A. Broring, S. Schmid, C.-K. Schindhelm, A. Khelil, S. Käbisch, D. Kramer, D. Le Phuoc, J. Mitic, D. Anicic, and E. Teniente, "Enabling IoT ecosystems through platform interoperability," *IEEE Softw.*, vol. 34, no. 1, pp. 54–61, Jan. 2017 [Online]. Available: http://ieeexplore.ieee.org/document/7819420/. [Accessed: 21-Feb-2019].

94. S. Kubler, J. Robert, A. Hefnawy, K. Framling, C. Cherifi, and A. Bouras, "Open IoT ecosystem for sporting event management," *IEEE Access*, vol. 5, pp. 7064–7079, 2017 [Online]. Available: http://ieeexplore.ieee.org/document/7898832/. [Accessed: 21-Feb-2019].

95. A. Karpenko, T. Kinnunen, M. Madhikermi, J. Robert, K. Främling, B. Dave, and A. Nurmine, "Data exchange interoperability in IoT ecosystem for smart parking and EV charging," *Sensors*, vol. 18, no. 12, p. 4404, Dec. 2018 [Online]. Available: http://www.mdpi.com/1424-8220/18/12/4404. [Accessed: 21-Feb-2019].

96. M. Elkhodr, S. Shahrestani, and H. Cheung, "The Internet of Things: New interoperability, management and security challenges," *Int. J. Netw. Secur. Appl.*, vol. 8, no. 2, pp. 85–102, Mar. 2016 [Online]. Available: http://www.aircconline.com/ijnsa/V8N2/8216ijnsa06.pdf. [Accessed: 22-Feb-2019].

97. M. Owayjan, B. Sleem, E. Saad, and A. Maroun, "Parking management system using mobile application," in *2017 Sensors Networks Smart and Emerging Technologies (SENSET)*, 2017, pp. 1–4.

98. S. Ravishankar and N. Theetharappan, "Cloud connected smart car park," in *2017 International Conference on I-SMAC (IoT in Social, Mobile, Analytics and Cloud) (I-SMAC)*, 2017, pp. 71–74.

99. N. Larisis, L. Perlepes, G. Stamoulis, and P. Kikiras, "Intelligent parking management system based on wireless sensor network technology," *Sens. Transducers*, vol. 18, pp. 100–112, Jan. 2013.

100. D. Kanteti, D. V. S. Srikar, and T. K. Ramesh, "Smart parking system for commercial stretch in cities," in *2017 International Conference on Communication and Signal Processing (ICCSP)*, 2017, pp. 1285–1289.

101. M. Y. I. Idris, E. M. Tamil, N. M. Noor, Z. Razak, and K. W. Fong, "Parking guidance system utilizing wireless sensor network and ultrasonic sensor," *Inf. Technol. J.*, vol. 8, no. 2, pp. 138–146, Feb. 2009 [Online]. Available: http://www.scialert.net/abstract/?doi=itj.2009.138.146. [Accessed: 24-Dec-2017].

102. S. Mathur, T. Jin, N. Kasturirangan, J. Chandrasekaran, W. Xue, M. Gruteser, and W. Trappe, "ParkNet: Drive-by sensing of road-side parking statistics," in *Proceedings of the 8th International Conference on Mobile Systems, Applications, and Services*, New York, NY, 2010, pp. 123–136.

103. S. Y. Cheung, S. C. Ergen, and P. Varaiya, "Traffic surveillance with wireless magnetic sensors," in *Proceedings of the 12th ITS World Congress*, 2005, vol. 1917, p. 173181.

104. S. Yoo, P. K. Chong, T. Kim, J. Kang, D. Kim, C. Shin, K. Sung, and B. Jang, "PGS: Parking guidance system based on wireless sensor network," in *2008 3rd International Symposium on Wireless Pervasive Computing*, 2008, pp. 218–222.

105. C. Trigona, B. Andò, V. Sinatra, C. Vacirca, E. Rossino, L. Palermo, S. Kurukunda, and S. Baglio, "Implementation and characterization of a smart parking system based on 3-axis magnetic sensors," in *2016 IEEE International Instrumentation and Measurement Technology Conference Proceedings*, 2016, pp. 1–6.

106. J. Chinrungrueng, S. Dumnin, and R. Pongthornseri, "iParking: A parking management framework," in *2011 11th International Conference on ITS Telecommunications*, 2011, pp. 63–68.

107. Z. Zhang, M. Tao, and H. Yuan, "A parking occupancy detection algorithm based on AMR sensor," *IEEE Sens. J.*, vol. 15, no. 2, pp. 1261–1269, Feb. 2015 [Online]. Available: http://ieeexplore.ieee.org/document/6919252/. [Accessed: 21-Feb-2019].

108. S. Banerjee, P. Choudekar, and M. K. Muju, "Real time car parking system using image processing," in *2011 3rd International Conference on Electronics Computer Technology*, 2011, vol. 2, pp. 99–103.

109. H.-C. Tan, J. Zhang, X.-C. Ye, H.-Z. Li, P. Zhu, and Q.-H. Zhao, "Intelligent car-searching system for large park," in *2009 International Conference on Machine Learning and Cybernetics*, 2009, vol. 6, pp. 3134–3138.

110. B. N. Silva, M. Khan, and K. Han, "Towards sustainable smart cities: A review of trends, architectures, components, and open challenges in smart cities," *Sustain. Cities Soc.*, vol. 38, pp. 697–713, Apr. 2018 [Online]. Available: https://linkinghub.elsevier.com/retrieve/pii/S2210670717311125. [Accessed: 24-Feb-2019].

111. S. Kubler, J. Robert, A. Hefnawy, C. Cherifi, A. Bouras, and K. Främling, "IoT-based smart parking system for sporting event management," in *Proceedings of the 13th International Conference on Mobile and Ubiquitous Systems: Computing, Networking and Services – MOBIQUITOUS 2016*, Hiroshima, Japan, 2016, pp. 104–114 [Online]. Available: http://dl.acm.org/citation.cfm?doid=2994374.2994390. [Accessed: 21-Feb-2019].

112. J. Rico, J. Sancho, B. Cendon, and M. Camus, "Parking easier by using context information of a smart city: Enabling fast search and management of parking resources," in *2013 27th International Conference on Advanced Information Networking and Applications Workshops*, Barcelona, 2013, pp. 1380–1385 [Online]. Available: http://ieeexplore.ieee.org/document/6550588/. [Accessed: 21-Feb-2019].

113. Z. Wei, Y. Li, Y. Zhang, and L. Cai, "Intelligent parking garage EV charging scheduling considering battery charging characteristic," *IEEE Trans. Ind. Electron.*, vol. 65, no. 3, pp. 2806–2816, Mar. 2018 [Online]. Available: http://ieeexplore.ieee.org/document/8012480/. [Accessed: 21-Feb-2019].

114. S. M. Bagher Sadati, J. Moshtagh, M. Shafie-khah, A. Rastgou, and J. P. S. Catalão, "Operational scheduling of a smart distribution system considering electric vehicles parking lot: A bi-level approach," *Int. J. Electr. Power Energy Syst.*, vol. 105, pp. 159–178, Feb. 2019 [Online]. Available: https://linkinghub.elsevier.com/retrieve/pii/ S0142061518304824. [Accessed: 21-Feb-2019].

115. M. S. Kuran, A. Carneiro Viana, L. Iannone, D. Kofman, G. Mermoud, and J. P. Vasseur, "A smart parking lot management system for scheduling the recharging of electric vehicles," *IEEE Trans. Smart Grid*, vol. 6, no. 6, pp. 2942–2953, Nov. 2015 [Online]. Available: http://ieeexplore.ieee.org/document/7056538/. [Accessed: 21-Feb-2019].

116. H. Y. Chang, H. W. Lin, Z. H. Hong, and T. L. Lin, "A novel algorithm for searching parking space in vehicle ad hoc networks," in *2014 Tenth International Conference on Intelligent Information Hiding and Multimedia Signal Processing*, 2014, pp. 686–689.

117. M. Santhiya, M. M. S. Karthick, and M. Keerthika, "Performance of various TCP in vehicular ad hoc network based on timer management," *Int. J. Adv. Res. Electr. Electron. Instrum. Energy*, vol. 2, no. 12, pp. 6160–6166, Dec. 2013 [Online]. Available: http://www.rroij.com/peer-reviewed/performance-of-various-tcp-invehicular-ad-hoc-networkbased-on-timer-management-42995.html. [Accessed: 26-Dec-2017].

118. W. Balzano and F. Vitale, "DiG-Park: A smart parking availability searching method using V2V/V2I and DGP-class problem," in *2017 31st International Conference on Advanced Information Networking and Applications Workshops (WAINA)*, 2017, pp. 698–703.

119. F. Al-Turjman, "Fog-based caching in software-defined information-centric networks," *Comput. Electr. Eng. J.*, vol. 69, no. 1, pp. 54–67, 2018.

120. C. Jeremiah and A. J. Nneka, "Issues and possibilities in vehicular ad-hoc networks (VANETs)," in *2015 International Conference on Computing, Control, Networking, Electronics and Embedded Systems Engineering (ICCNEEE)*, 2015, pp. 254–259.

121. X. Li, M. C. Chuah, and S. Bhattacharya, "UAV assisted smart parking solution," in *2017 International Conference on Unmanned Aircraft Systems (ICUAS)*, Miami, FL, 2017, pp. 1006–1013 [Online]. Available: http://ieeexplore.ieee.org/document/7991353/. [Accessed: 21-Feb-2019].

122. M. D'Aloia, M. Rizzi, R. Russo, M. Notarnicola, and L. Pellicani, "A marker-based image processing method for detecting available parking slots from UAVs," in *New Trends in Image Analysis and Processing – ICIAP 2015 Workshops*, vol. 9281, V. Murino, E. Puppo, D. Sona, M. Cristani, and C. Sansone, Eds. Cham: Springer International Publishing, 2015, pp. 275–281 [Online]. Available: http://link.springer. com/10.1007/978-3-319-23222-5_34. [Accessed: 21-Feb-2019].

123. "Streetline," *Streetline* [Online]. Available: https://www.streetline.com/. [Accessed: 26-Dec-2017].

124. C. H. Huang, H. S. Hsu, H. R. Wang, T. Y. Yang, and C. M. Huang, "Design and management of an intelligent parking lot system by multiple camera platforms," in *2015 IEEE 12th International Conference on Networking, Sensing and Control*, 2015, pp. 354–359.

125. F. Al-Turjman and S. Alturjman, "5G/IoT-enabled UAVs for multimedia delivery in industry-oriented applications," *Multimedia Tools Appl. J.*, 2018. doi: 10.1007/s11042-018-6288-7.

126. F. Al-Turjman, "QoS–aware data delivery framework for safety-inspired multimedia in integrated vehicular-IoT," *Comput. Commun. J.*, vol. 121, pp. 33–43, 2018.

127. P. Lee, H.-P. Tan, and M. Han, "Demo: A solar-powered wireless parking guidance system for outdoor car parks," in *Proceedings of the 9th ACM Conference on Embedded Networked Sensor Systems*, New York, NY, 2011, pp. 423–424.

128. F. Al-Turjman and S. Alturjman, "Context-sensitive access in industrial Internet of Things (IIoT) healthcare applications," *IEEE Trans. Indus. Infor.*, vol. 14, no. 6, pp. 2736–2744, 2018.

10 Correctness of an Authentication Scheme for Managing Demand Response in Smart Grid

Shehzad Ashraf Chaudhry and Khalid Yahya
Istanbul Gelisim University

Fadi Al-Turjman
Near East University

CONTENTS

10.1 INTRODUCTION

The smart grid (SG) unifies the information and communication technologies and the traditional power generation grid to provide scuffle-free and convenient method for demand response management in power sector. SG empowers the optimized and customized flow of energy between the utility controllers and the consumers, and extends advantage to both the generation sources and the consumers owing to realization of real-time consumption. The SG manages demand response bidirectional communication among the smart meters and the utility controls (UCs), and the former is responsible for monitoring and requesting the real-time power flow from the latter. Typically, this management takes place every 10–15 minutes through bidirectional communication among both. The communication can be realized through various technologies including short- and medium- to long-range communication methods [1,2]. The DR management in SG environment has become a critical concern due to the lack of security in ICT-embedded framework which needs concerted efforts to be fixed. According to the U.S. Department of Energy (DoE), the demand of electricity has augmented by 30% since 1988, and it is further anticipated to rise with nearly the same pace in the approaching decades. Nevertheless, the development of state-of-the-art smart grid systems, 15% as reported, could not keep pace with the increasing demand. Despite many SG-based initiated projects, as reported by DoE, it is still a challenging task to effectively control the SG-based demand response management. Cyber physical social systems (CPSSs) refer to the systems embedded with physical as well as software computational resources [3]. The CPSS may be centrally managed by the controllers using intelligent algorithms. The smart grid can be rightly regarded as a CPSS due to involvement of human element or social space, while for being CPSS, the SG system can produce efficient and reliable data analytics. This emphasizes the need to incorporate more and more physical entities bearing ICT-based intelligence with social interactions in order to produce more insightful information and results [4]. In SG, bulk of the data may be produced on hourly basis from advanced metering infrastructure bearing sensors, which needs to be handled with CPSS to improve the smart grid services. A typical SG infrastructure is shown in Figure 10.1, where the smart meters attached with different types of consumers act for demand through communicating with UCs connected with power generation centers; the UC is responsible for responding accordingly with appropriate action regarding the power flow. Both these entities make the real-time demand response management. As the whole SG infrastructure is relying on insecure public network, the communication between several devices can cause severe threats. Any attacker can easily intercept the demand or response messages and can also initiate other attacks to expose user-related secret credentials. The attacker can also fluctuate the power supply by deceiving the grid infrastructure. Such attacker initiatives/capabilities call for an

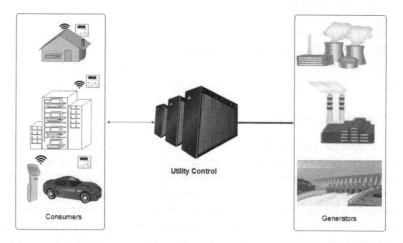

FIGURE 10.1   Demand response management diagram.

authentication scheme to secure the communication between different grid entities. To address those security issues, there must be robust communication infrastructure in the form of authentication protocols, supporting secure information exchange among the legitimate entities, and maintaining the privacy as well.

It is not been so long that the smart grid security got designers attention; however, many authentication protocols were proposed in recent time. In 2016, Tsai and Lo [5] proposed a pairing- and identity-based authentication scheme for securing smart grid communication. The scheme realized the need of privacy and the direct device-to-device authentication without intervention of third party. Odelu et al. [6] drew designers' attention towards inherent problem of the insecurity of the scheme of Tsai and Lo against the leakage of particular session parameters, which can expose the privacy of meters. To counter the drawback of Ref. [5], Odelu et al. presented another scheme based on pairing and identities. However, their scheme lacks initial verification of request message, which can invite an adversary to send forged messages and receive responses. The adversary may never succeed in masquerading any smart meter, as the forging is exposed in second message from the adversary (the response of the challenge). However, a large number of received forged messages may affect the quality of service. Another security scheme was presented by Doh et al. [7] to secure the communication among smart meter and the UC. Later to bridge some gaps, Saxena et al. [8] designed yet another scheme. He et al. [9] also presented an elliptic curve cryptography (ECC)-based scheme for securing smart grid communication as well as to provide user anonymity. Due to avoidance of pairing and modular exponentiation, the scheme of He et al. extended computational and communicative efficiencies. Later in 2017, Bayat et al. [10] offered another efficient scheme using chaotic maps. Mohammadali et al. [11] also proposed an ECC-based mechanism for smart grid security. Arguing based on the proof of Mahmood et al. [12], the scheme of Mohammadali et al. was insecure against exposure of private key related with trusted third part. Another ECC-based scheme of Mahmood et al. [13] was proved as insecure by Mood and Nikooghadam [14]. As per Ref. [14], the scheme of Mahmood et al. [12] was also lacking forward secrecy. In 2020, Liang et al. revealed that another scheme of Mahmood et al. [15] cannot face exposure of ephemeral secrets. In other effort, Challa et al. [16] put forward another scheme. Nevertheless, Chaudhry et al. [17] pointed out a severe design flaw and argued that the scheme cannot be deployed in real-world scenario owing to the mentioned flaw. The improved scheme of Chaudhry et al. does not provide device-to-device (D2D) communication and can extend authentication among communicating entities through some trusted intervening party.

10.1.1 Motivations and Contributions

To counter the discussed drawbacks, in 2019, Kumar et al. [18] proposed an ECC-based authentication scheme *ECCAuth* to establish a mutually agreed session key between SG device and UC for secure communication as well as smooth performance of SG operations and in return controlling the demand response in SG-based systems. However, in this study, we intend to prove that *ECCAuth* presented by Kumar et al. is not practical and has incorrect login and authentication phase, leading towards complete inapplicability in real-world SG scenarios. Moreover, in *ECCAuth*, there is no initial verification on UC side, and any adversary can force UC to process the illegal requests. The research contributions are illustrated in Table 10.1.

1. We provided an over view of recent progress for securing the smart grid.
2. We review and cryptanalyze a very recent scheme *ECCAuth* by Kumar et al.
3. We expose a critical design flaw, which can restrict the deployment of *ECCAuth* in real-world scenarios.

10.2 REVIEW OF THE SCHEME OF KUMAR et al.

This section presents a brief review of *ECCAuth* proposed by Kumar et al. *ECCAuth* involves three participants: (1) $\mathcal{TA}$, (2) utility control UC_b, and (3) smart device SD_a. All the smart devices SD_a: $\{a = 1, 2, 3, \ldots l\}$ and utility control devices UC_b: $\{b = 1, 2, 3, \ldots m\}$, registered with an SG device SD_a can get authenticated from its corresponding UC_b. The details of the phases involving *ECCAuth* processes are given next.

10.2.1 SYSTEM SETUP

To accomplish the setting up of the system, $\mathcal{TA}$ selects an elliptic curve $E_p(\alpha, \beta)$ over finite field Z_p along with a base point $G \in E_p(\alpha, \beta)$ of large order n s.t. $n.G = O$ (a point at infinity). The p is selected as a very large prime number satisfying $4\alpha^3 - 27\beta^2 \neq 0$ $mod\ p$. $\mathcal{TA}$ then selects x as private and $Q = xG$ as its own public key. $\mathcal{TA}$ also selects a secure one-way function $h(.)$ and finally publishes $\{E_p(\alpha, \beta), G, Q, h(.)\}$.

10.2.2 SG DEVICE REGISTRATION

For registering, each SG device SD_a: $\{i = 1, 2, \ldots m\}$ selects unique ID_a and computes $RID_a = h(ID_a\|x)$. $\mathcal{TA}$ then using registration timestamp RTS_a further computes $TC_a = h(x\|RTS_a)$. and finally stores $\{RID_a, TC_a, E_p(\alpha, \beta), G, Q, h(.)\}$ in the memory of SD_a.

TABLE 10.1
Notation Guide

Notations	Description
SD_a, ID_a	Smart device, SD_a's identity
$\mathcal{TA}, \mathcal{A}$	Trusted authority, adversary
RTS_k, RTS_b	Registration timestamp of entity k
p, Z_p	Large prime, p's finite field
$G, q.G$	$E_p(\alpha, \beta)$ point, scalar multiplier
$\Delta T, \|$	Delay tolerance, concatenation
UC_b, ID_b	Utility control, SD_a's identity
RID_k	Pseudo identity of entity k
T1, T2, T3	Several timestamps
$E_p(\alpha, \beta), \oplus$	Elliptic curve, XOR operator
$x, Q = x.G$	Private, public key pair of $\mathcal{TA}$
$h(.), \overset{?}{=}$	Hash, relational equality

10.2.3 UC Registration

For registering each UC_b: $\{j = 1, 2, \dots n\}$, $\mathcal{TA}$ selects unique ID_b and computes $RID_b = h(ID_b \| x)$. $\mathcal{TA}$ then using registration timestamp $RT\,S_b$ further computes $TC_a = h(x \| RT\,S_b)$ and finally stores $\{RID_b, TC_b, G, Q, RID_a$: $\{a = 1, 2, \dots m\}, h(.), E_p(\alpha, \beta)\}$ in the memory of UC_b.

10.2.4 Authentication

In Kumar et al.'s scheme, SD_a initiates authentication phase to furnish a secure session key with UC_b. The steps as illustrated in Figure 10.2 and briefed next are performed between SD_a and UC_b to complete this phase:

KDA 1: $SD_a \rightarrow UC_b : \{m_1\}$

SD_a selects $u \in Z_p^*$ randomly and generates current timestamp T_1. SD_a then computes $\mathrm{Ua} = u.\,G$ and $C_a = h(TC_a \| T_1) \oplus h(RID_a \| U_a \| T_1)$. Finally, SD_a sends $m_1 = \{U_a, C_a, T_1\}$ to UC_b.

KDA 2: $UC_b \rightarrow SD_a : \{m_2\}$

UC_b after receiving m_1 first verifies message freshness by checking $|T_1 - T_1^*| \leq 0$, and upon success UC_b computes $D_b = C_a \oplus h\,(RID_a \| U_a \| T_1)$. UC_b then selects $v \in Z_p^*$ and new timestamp T_2 and computes $V_b = v.G$, $W_b = v.U_a = (uv).G$, $SK_{ab} = h(W_b \| D_b \| h(RID_b \| TC_b \| T_2))$, $SKV_{ab} = h(SK_{ab} \| RID_b \| T_2)$ and $Z_b = h(RID_b \| TC_b \| T_2) \oplus h(RID_a \| U_a \| V_a \| T_2)$. UC_b completes this step by sending $m_2 = \{V_b, Z_b, SKV_{ab}, T_2\}$ to SD_a.

KDA 3: $SD_a \rightarrow UC_b : \{m_3\}$

SD_a after receiving m_2 first verifies message freshness by checking $|T_2 - T_2^*| \leq 0$, and upon success SD_a computes $E_a = Z_b \oplus h(RID_a \| U_a \| V_a \| T_2)$, $W_a' = u.V_b = (uv).G$, $SK_{ab}' = h\left(W_a' \| h(TC_a \| T_1) \| E_a\right)$ and $SKV_{ab}' = h\left(SK_{ab}' \| RID_a \| T_a\right)$.

SD_a then checks the equality $SKV_{ab}' \overset{?}{=} SKV_{ab}$, and upon success SD_a generates T_3 and computes $SKV_{ab}^* = h\left(SK_{ab}' \| RID_a \| V_b \| T_3\right)$. To complete this step, SD_a sends $m_3 = \left\{SKV_{ab}^* \cdot T_3\right\}$ to UC_b.

KDA 4: UC_b after receiving m_3 first verifies message freshness by checking $|T_3 - T_3^*| \leq 0$, and upon success UC_b computes $SKV_{ab}^{**} = h\left(SK_{ab} \| RID_a \| V_b \| T_3\right)$ and compares $SKV_{ab}^* \overset{?}{=} SKV_{ab}^{**}$, on success UC_b considers SD_a as legal and authenticated device.

10.2.5 SG Device Dynamic Addition

The dynamic addition of a new device SD_a^{new} requires very similar procedure as of SG device registration. For dynamic addition of a device SD_a^{new}, $\mathcal{TA}$ selects unique ID_a^{new} and computes $\left\{RID_a^{new}, TC_a^{new}, E_p(\alpha, \beta), G, Q, h(.)\right\}$. $\mathcal{TA}$ then using registration timestamp RTS_a^{new} further computes $TC_a^{new} = h\left(x \| RTS_a^{new}\right)$, then stores $\left\{RID_a^{new}, TC_a^{new}, E_p(\alpha, \beta), G, Q, h(.)\right\}$ in the memory of SD_a^{new}, and deploys it in the system. $\mathcal{TA}$ finally sends RID_a^{new} to each UC_b.

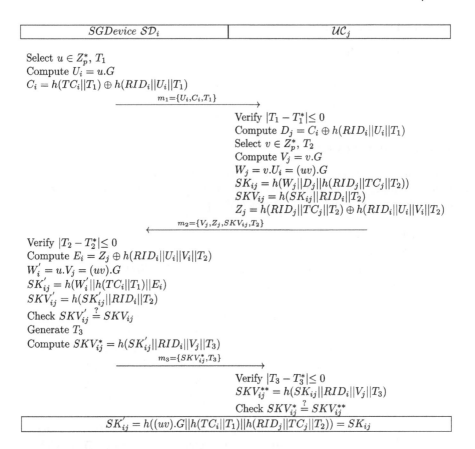

$SGDevice\ SD_i$	UC_j

Select $u \in Z_p^*$, T_1
Compute $U_i = u.G$
$C_i = h(TC_i\|T_1) \oplus h(RID_i\|U_i\|T_1)$

$$m_1=\{U_i,C_i,T_1\} \longrightarrow$$

                                          Verify $|T_1 - T_1^*| \leq 0$
                                          Compute $D_j = C_i \oplus h(RID_i\|U_i\|T_1)$
                                          Select $v \in Z_p^*$, T_2
                                          Compute $V_j = v.G$
                                          $W_j = v.U_i = (uv).G$
                                          $SK_{ij} = h(W_j\|D_j\|h(RID_j\|TC_j\|T_2))$
                                          $SKV_{ij} = h(SK_{ij}\|RID_i\|T_2)$
                                          $Z_j = h(RID_j\|TC_j\|T_2) \oplus h(RID_i\|U_i\|V_i\|T_2)$

$$\longleftarrow m_2=\{V_j,Z_j,SKV_{ij},T_2\}$$

Verify $|T_2 - T_2^*| \leq 0$
Compute $E_i = Z_j \oplus h(RID_i\|U_i\|V_i\|T_2)$
$W_i' = u.V_j = (uv).G$
$SK_{ij}' = h(W_i'\|h(TC_i\|T_1)\|E_i)$
$SKV_{ij}' = h(SK_{ij}'\|RID_i\|T_2)$
Check $SKV_{ij}' \overset{?}{=} SKV_{ij}$
Generate T_3
Compute $SKV_{ij}^* = h(SK_{ij}'\|RID_i\|V_j\|T_3)$

$$m_3=\{SKV_{ij}^*,T_3\} \longrightarrow$$

                                          Verify $|T_3 - T_3^*| \leq 0$
                                          $SKV_{ij}^{**} = h(SK_{ij}\|RID_i\|V_j\|T_3)$
                                          Check $SKV_{ij}^* \overset{?}{=} SKV_{ij}^{**}$

$SK_{ij}' = h((uv).G\|h(TC_i\|T_1)\|h(RID_j\|TC_j\|T_2)) = SK_{ij}$

FIGURE 10.2   Kumar et al.'s authentication procedure.

10.2.6   UC DYNAMIC ADDITION

The dynamic addition of a utility control device UC_b^{new} requires very similar procedure as of SG device registration. For dynamic addition of a new UC_b^{new}, $\mathcal{TA}$ selects unique ID_b^{new} and computes $RID_b^{new} = h\left(ID_b^{new} \| x\right)$. $\mathcal{TA}$ then using registration timestamp RTS_b^{new}. $\mathcal{TA}$ further computes $TC_b^{new} = h\left(x \| RTS_b^{new}\right)$. $\mathcal{TA}$ stores $\left\{RID_b^{new}, TC_b^{new}, G, Q, RID_a : \{i=1,2,...m\}, h(.), E_p(\alpha,\beta)\right\}$ in the memory of UC_b. Finally, it broadcasts the message to inform all entities in the grid environment about the addition of UC_b^{new}.

10.3   DISADVANTAGES OF KUMAR ET AL.'S SCHEME

The disadvantages of Kumar et al.'s scheme are explored in this section. In following subsections, it is shown that the scheme of Kumar et al. is having correctness issues, leading to inapplicability of their scheme in real-world SG environments. Moreover, the lack of verification on $\mathcal{UC}_b$ side after receiving initial request message may burden the system by forcing the UC_b to process an illegal request by using only fresh timestamp by any active adversary. Such disadvantage can also lead to denial of service (DoS) and battery flattening attacks. Following subsections describe these disadvantages briefly.

10.3.1 INCORRECTNESS

Here, we show that the scheme of Kumar et al. is unable to extend authentication between two system entities (i.e., UC_b and SD_a). In Kumar et al.'s scheme, SD_a after sending request may never complete the process and get a reply from UC_b. The situation can be simulated as follows:

1. SD_a generates T_1 and computes

$$U_a = u.G \tag{10.1}$$

$$C_a = h(TC_a \parallel T_1) \oplus h(RID_a \parallel U_a \parallel T_1) \tag{10.2}$$

   SD_a then sends $m_1 = \{U_a, C_a, T_1\}$ to UC_b.
2. UC_b upon reception of m_1 verifies the timestamp freshness and then computes

$$D_b = C_a \oplus h(RID_a \parallel U_a \parallel T_1) \tag{10.3}$$

3. UC_b computes $D_b = h(TC_a \parallel T_1)$ through Eq. (10.3) which requires to perform an exclusive or of received C_a with $h(RID_a \parallel U_a \parallel T_1)$. SD_a sent T_1 and U_a in message m_1 to UC_b. However, UC_b does not know the user-specific parameter RID_a. UC_b stores a verifier table of entries $\{RID_a | i = 1,2...l\}$ where each RID_a corresponds to a specific smart device say SD_a. Now, there are many devices (say l) attached with UC_b; whenever, UC_b receives an authentication request, it needs to extract user-specific RID_a. However, in the request message, SD_a does not send any clue about itself. None of the received parameters $(m_1 = \{U_a, C_a, T_1\})$ ($m_1 = U_a$, C_a, T_1) is helpful to know the identity of requesting user. Therefore, UC_b is unable to determine which $\{RID_a | a = 1,2...l\}$ out of l stored pseudo identities it has to use for computation of D_b in Eq. (10.3). Hence, scheme will not proceed further, and a halt situation will occur. The process cannot complete normally.

10.3.2 NO INITIAL VERIFICATION ON UC_b SIDE

On receiving m_1, UC_b does not perform any request verification except verification of timestamp freshness. The verification on UC_b side takes place in the second message m_3 by SD_a. Any adversary can randomly generate U_a, C_a along with current timestamp T_a and can send $\{U_a, C_a, T_a\}$ to UC_b. Such request message may force UC_b to process the request and send reply message. However, UC_b is able to abort the session, when it receives m_3. However, the lack of verification on UC_b side may lead to unnecessary computation and communication operations, and a large number of such application, especially in wireless environment can lead to battery flattening and/or DoS attack.

10.4   CONCLUSION

In this chapter, we analyzed a recent scheme *ECCAuth* proposed to extend a secure session among communicating entities in smart grid environment. However, this chapter clarified that *ECCAuth* has serious design flaws to work with multiple devices. The analysis has also shown that the lack of initial verification on receiving side may force an attacker to initiate forged messages and receive responses, which can badly effect the efficiency. This chapter aimed to show that besides security, the scheme correctness should be verified, as if the scheme cannot be deployed in real-world scenario, the scheme's robustness may not benefit the scientific and otherwise community.

REFERENCES

1. V. C. Gungor, D. Sahin, T. Kocak, S. Ergut, C. Buccella, C. Cecati, G. P. Hancke, Smart grid technologies: Communication technologies and standards, *IEEE Transactions on Industrial Informatics* 7 (4), 529–539 (2011).
2. A. Metke, R. Ekl, Security technology for smart grid networks, *IEEE Transactions on Smart Grid* 1 (1), 99–107 (2010).
3. J. Zeng, L. T. Yang, M. Lin, H. Ning, A. S. J. Ma, Cyberphysical-social systems and their system-level design methodology, *Future Generation Computer Systems* (2016). doi:10.1016/j.future.2016.06.034.
4. Q. Jing, A. V. Vasilakos, J. Wan, J. Lu, D. Qiu, Security of the Internet of Things: Perspectives and challenges, *Wireless Networks* 20 (8), 2481–2501 (2014).
5. J.-L. Tsai, N.-W. Lo, Secure anonymous key distribution scheme for smart grid, *IEEE Transactions on Smart Grid* 7 (2), 906–914 (2016).
6. V. Odelu, A. K. Das, M. Wazid, M. Conti, Provably secure authenticated key agreement scheme for smart grid, *IEEE Transactions on Smart Grid* (2016). doi:10.1109/TSG.2016.2602282.
7. I. Doh, J. Lim, K. Chae, Secure authentication for structured smart grid system, in: *International Conference on Innovative Mobile and Internet Services in Ubiquitous Computing (IMIS-15)*, Fukuoka, Japan, pp. 200–204 (2015).
8. N. Saxena, B. J. Choi, R. Lu, Authentication and authorization scheme for various user roles and devices in smart grid, *IEEE Transactions on Information Forensics and Security* 11 (5), 907–921 (2016).
9. D. He, H. Wang, M. K. Khan, L. Wang, Lightweight anonymous key distribution scheme for smart grid using elliptic curve cryptography, *IET Communications* 10 (14), 1795–1802 (2016).
10. M. Bayat, M. B. Atashgah, M. R. Aref, A secure and efficient chaotic maps based authenticated key-exchange protocol for smart grid, *Wireless Personal Communications* 97 (2), 2551–2579 (2017).
11. A. Mohammadali, M. S. Haghighi, M. H. Tadayon, A. Mohammadi-Nodooshan, A novel identity-based key establishment method for advanced metering infrastructure in smart grid, *IEEE Transactions on Smart Grid* 9 (4), 2834–2842 (2018).
12. K. Mahmood, J. Arshad, S. A. Chaudhry, S. Kumari, An enhanced anonymous identity-based key agreement protocol for smart grid advanced metering infrastructure, *International Journal of Communication Systems* 32 (16), e4137 (2019).
13. K. Mahmood, S. A. Chaudhry, H. Naqvi, S. Kumari, X. Li, A. K. Sangaiah, An elliptic curve cryptography based lightweight authentication scheme for smart grid communication, *Future Generation Computer Systems* 81, 557–565 (2018).

14. D. Abbasinezhad-Mood, M. Nikooghadam, Design and hardware implementation of a security-enhanced elliptic curve cryptography based lightweight authentication scheme for smart grid communications, *Future Generation Computer Systems* 84, 47–57 (2018).

15. K. Mahmood, X. Li, S. A. Chaudhry, H. Naqvi, S. Kumari, A. K. Sangaiah, J. J. Rodrigues, Pairing based anonymous and secure key agreement protocol for smart grid edge computing infrastructure, *Future Generation Computer Systems* 88, 491–500 (2018).

16. S. Challa, A. K. Das, P. Gope, N. Kumar, F. Wu, E. Yoon, A. V. Vasilakos, Design and analysis of authenticated key agreement scheme in cloud-assisted cyber-physical systems, *Future Generation Computer Systems* (2018). doi:10.1016/j.future.2018.04.019.

17. S. A. Chaudhry, T. Shon, F. Al-Turjman, M. H. Alsharif, Correcting design flaws: An improved and cloud assisted key agreement scheme in cyber physical systems, *Computer Communications* 153, 527–537 (2020). doi:10.1016/j.comcom.2020.02.025.

18. N. Kumar, G. S. Aujla, A. K. Das, M. Conti, Eccauth: A secure authentication protocol for demand response management in a smart grid system, *IEEE Transactions on Industrial Informatics* 15 (12), 6572–6582 (2019). doi:10.1109/TII.2019.2922697.

11 An Overview about the Cyberattacks in Grid and Like Systems

Fadi Al-Turjman and Ramiz Salama
Near East University

CONTENTS

11.1 INTRODUCTION

A cyberattack is an intentional exploitation of computer systems, networks, and technology-dependent enterprises. These attacks use malicious code to modify computer code, data, or logic, culminating into destructive consequences that can compromise your data and promulgate cybercrimes such as information and identity theft. A cyberattack is also known as a computer network attack (CNA).

The technological innovation of cybersystems and increased dependence of individuals, societies, and nations on them have brought new, real, and ever-changing threat landscapes. In fact, the threats evolve faster than they can be assessed. The technological innovation that brought ease and efficiency to our lives has been met by similar innovation to take advantage of cybersystems for other gains. More threat actors are noted to be sponsored by nation-states, and the skills and capabilities of organizations to defend against these attacks are lagging. This warrants an increase in automation of threat analysis and response as well as increased adoption of security measures by at-risk organizations. Thus, to properly prepare defenses and mitigations to the threats introduced by cyber, it is necessary to understand these threats. Accordingly, this chapter provides overview of top cybersecurity threats together with current and emerging trends. The analyses include general trends in the complexity of attacks, actors, and the maturity of skills and capabilities of organizations to defend against attacks. Top threats are discussed with regard to instances of attacks and strategies for mitigation within the kill chain. A brief discussion of threat agents and attack vectors adds context to the threats.

11.2 THREATS TO CYBERSECURITY

11.2.1 PHISHING

Phishing is the practice of sending fraudulent emails that resemble emails from reputable sources. The aim is to steal sensitive data like credit card numbers and login information, or to install malware on the victim's machine. It's the most common type of cyberattack. You can help protect yourself through education or a technology solution that filters malicious emails.

11.2.1.1 Phishing Tips
- Avoid strangers, and check name and email address.
- Don't rush, and be suspicious of emails marked "urgent."

- Notice mistakes in spelling and grammar.
- Beware of generic greetings, "dear sir/ma'am."
- Don't be lured by incredible "deals."
- Hover over the link before you click to ensure it has a secure URL (https://).
- Never give out personal or financial information based on an email request.
- Don't trust links or attachments in unsolicited emails.

11.2.1.2   How Does Phishing Work?

Phishing starts with a fraudulent email or other communication that is designed to lure a victim. The message is made to look as though it comes from a trusted sender. If it fools the victim, he or she is coaxed into providing confidential information, often on a scam website. Sometimes, malware is also downloaded onto the target's computer.

11.2.1.3   What Are the Dangers of Phishing Attacks?

Sometimes, attackers are satisfied with getting a victim's credit card information or other personal data for financial gain. Other times, phishing emails are sent to obtain employee login information or other details for use in an advanced attack against a specific company. Cybercrime attacks such as advanced persistent threats (APTs) and ransomware often start with phishing.

11.2.1.4   How Do I Protect against Phishing Attacks?

11.2.1.4.1   User Education

One way to protect your organization from phishing is user education. Education should involve all employees. High-level executives are often a target. Teach them how to recognize a phishing email and what to do when they receive one. Simulation exercises are also a key for assessing how your employees react to a staged phishing attack.

11.2.1.4.2   Security Technology

No single cybersecurity technology can prevent phishing attacks.

11.2.1.5   Examples of Phishing Attacks

11.2.1.5.1   Spear Phishing

Spear phishing targets specific individuals instead of a wide group of people. Attackers often research their victims on social media and other sites. That way, they can customize their communications and appear more authentic. Spear phishing is often the first step used to penetrate a company's defenses and carry out a targeted attack. According to the SANS Institute, 95% of all attacks on enterprise networks are the result of successful spear phishing.

11.2.1.5.2   Whaling

When attackers go after a "big fish" like a CEO, it's called whaling. These attackers often spend considerable time profiling the target to find the opportune moment and means of stealing login credentials. Whaling is of particular concern because high-level executives are able to access a great deal of company information.

11.2.1.5.3   Pharming

Similar to phishing, pharming sends users to a fraudulent website that appears to be legitimate. However, in this case, victims do not even have to click a malicious link to be taken to the bogus site. Attackers can infect either the user's computer or the website's DNS server and redirect the user to a fake site even if the correct URL is typed in.

11.2.1.5.4   Deceptive Phishing

Deceptive phishing is the most common type of phishing. In this case, an attacker attempts to obtain confidential information from the victims. Attackers use the information to steal money or to launch other attacks. A fake email from a bank asking you to click a link and verify your account details is an example of deceptive phishing.

11.2.1.5.5   Office 365 Phishing

The methods used by attackers to gain access to an Office 365 email account are fairly simple and becoming the most common. These phishing campaigns usually take the form of a fake email from Microsoft. The email contains a request to log in, stating the user needs to reset their password and hasn't logged in recently, or that there's a problem with the account that needs their attention. A URL is included, enticing the user to click to remedy the issue.

11.2.1.6   Advanced Email Security Protection

Attackers rely primarily on email to distribute spam, malware, and other threats. To prevent breaches, you need a powerful email security solution.

11.2.2   Ransomware

Ransomware is a type of malicious software. It is designed to extort money by blocking access to files or the computer system until the ransom is paid. Paying the ransom does not guarantee that the files will be recovered or the system restored.

Ransomware is a type of malicious software, also known as malware. It encrypts a victim's data until the attacker is paid a predetermined ransom. Typically, the attacker demands payment in a form of cryptocurrency such as Bitcoin. Only then will the attacker send a decryption key to release the victim's data.

A number of ransomware variants have appeared in recent years, which we'll describe in greater detail next. We will also explain how you can protect your system against future attacks.

11.2.2.1   How Does Ransomware Work?

Ransomware is typically distributed through a few main avenues. These include email phishing, malvertising (malicious advertising), and exploit kits. After it is distributed, the ransomware encrypts selected files and notifies the victim of the required payment. Watch demo of ransomware attack.

11.2.2.2 Ten Ways to Protect Yourself from Ransomware

11.2.2.2.1 Back Up All Your Data

In the event of an attack, you can power down the endpoint, reimage it, and reinstall your recent backup. You'll have all your data, and you'll prevent the ransomware from spreading to other systems.

11.2.2.2.2 Patch Your Systems

Make a habit of updating your software regularly. Patching commonly exploited third-party software will foil many attacks.

11.2.2.2.3 Educate Users on Attack Sources

The weakest link in the security chain is usually human. Educate your users on whom and what to trust. Empower them not to fall for phishing or other schemes.

11.2.2.2.4 Protect Your Network

Take a layered approach, with security infused from the endpoint to email to the DNS layer. Use technologies such as a next-generation firewall (NGFW) or an intrusion prevention system (IPS).

11.2.2.2.5 Segment Network Access

Limit the resources that an attacker can access. By dynamically controlling access at all times, you help ensure that your entire network is not compromised in a single attack.

11.2.2.2.6 Keep a Close Eye on Network Activity

Being able to see everything happening across your network and data center can help you uncover attacks that bypass the perimeter. Deploy a demilitarized zone (DMZ) or add a layer of security to your local area network (LAN).

11.2.2.2.7 Prevent Initial Infiltration

Most ransomware infections occur through an email attachment or a malicious download. Diligently block malicious websites, emails, and attachments through a layered security approach and a company-sanctioned file-sharing program.

11.2.2.2.8 Arm Your Endpoints

Antivirus solutions on your endpoints don't suffice anymore. Set up privileges, so they perform tasks such as granting the appropriate network shares or user permissions on endpoints. Two-factor authentications will also help.

11.2.2.2.9 Gain Real-Time Threat Intelligence

Know your enemy. Take advantage of threat intelligence from organizations such as Talos to understand security information and emerging cybersecurity threats.

11.2.2.2.10 Say No to Ransom

Never, ever pay the ransom. There's no guarantee you'll get your data back, and you're only fueling the cybercriminals for more attacks.

11.2.3 Malware

Malware is a type of software designed to gain unauthorized access or to cause damage to a computer. Malware is intrusive software that is designed to damage and destroy computers and computer systems. Malware is a contraction for "malicious software." Examples of common malware include viruses, worms, Trojan viruses, spyware, adware, and ransomware.

11.2.3.1 How Do I Protect My Network Against Malware?

Typically, businesses focus on preventative tools to stop breaches. By securing the perimeter, businesses assume they are safe. Some advanced malware, however, will eventually make their way into your network. As a result, it is crucial to deploy technologies that continually monitor and detect malware that has evaded perimeter defenses. Sufficient advanced malware protection requires multiple layers of safeguards along with high-level network visibility and intelligence.

11.2.3.2 How Do I Detect and Respond to Malware?

Malware will inevitably penetrate your network. You must have defenses that provide significant visibility and breach detection. In order to remove malware, you must be able to identify malicious actors quickly. This requires constant network scanning. Once the threat is identified, you must remove the malware from your network. Today's antivirus products are not enough to protect against advanced cyberthreats.

11.2.3.3 Types of Malware

11.2.3.3.1 Virus

Viruses are a subgroup of malware. A virus is malicious software attached to a document or file that supports macros to execute its code and spread from host to host. Once downloaded, the virus will lay dormant until the file is opened and in use. Viruses are designed to disrupt a system's ability to operate. As a result, viruses can cause significant operational issues and data loss.

11.2.3.3.2 Worms

Worms are a malicious software that rapidly replicates and spreads to any device within the network. Unlike viruses, worms do not need host programs to disseminate. A worm infects a device via a downloaded file or a network connection before it multiplies and disperses at an exponential rate. Like viruses, worms can severely disrupt the operations of a device and cause data loss.

11.2.3.3.3 Trojan Virus

Trojan viruses are disguised as helpful software programs. But once the user downloads it, the Trojan virus can gain access to sensitive data and then modify, block, or delete the data. This can be extremely harmful to the performance of the device. Unlike normal viruses and worms, Trojan viruses are not designed to self-replicate.

11.2.3.3.4   Spyware

Spyware is malicious software that runs secretly on a computer and reports back to a remote user. Rather than simply disrupting a device's operations, spyware targets sensitive information and can grant remote access to predators. Spyware is often used to steal financial or personal information. A specific type of spyware is a keylogger, which records your keystrokes to reveal passwords and personal information.

11.2.3.3.5   Adware

Adware is malicious software used to collect data on your computer usage and provide appropriate advertisements to you. While adware is not always dangerous, in some cases adware can cause issues for your system. Adware can redirect your browser to unsafe sites and can even contain Trojan horses and spyware. Additionally, significant levels of adware can slow down your system noticeably. Because not all adware is malicious, it is important to have protection that constantly and intelligently scans these programs.

11.2.3.3.6   Ransomware

Ransomware is malicious software that gains access to sensitive information within a system, encrypts that information so that the user cannot access it, and then demands a financial payout for the data to be released. It is commonly part of a phishing scam. By clicking a disguised link, the user downloads the ransomware. The attacker proceeds to encrypt specific information that can only be opened by a mathematical key they know. When the attacker receives payment, the data is unlocked.

11.2.3.3.7   Fileless Malware

Fileless malware is a type of memory-resident malware. As the term suggests, it is malware that operates from a victim's computer's memory, not from files on the hard drive. Because there are no files to scan, it is harder to detect than traditional malware. It also makes forensics more difficult because the malware disappears when the victim computer is rebooted. In late 2017, the Cisco Talos threat intelligence team posted an example of fileless malware that they called DNSMessenger.

11.3   EVOLUTIONS IN THE THREAT LANDSCAPE

11.3.1   THE TOP FIFTEEN THREATS IN 2018

In order, the top fifteen threats in 2018 were malware, web-based attacks, web application attacks, phishing, denial of service (DoS), spam, botnets, data breaches, insider threat, physical manipulation/ damage/theft/loss, information leakage, identity theft, cryptojacking, ransomware, and crypto espionage. The annual change in ranking of each cyberthreat since 2012 is summarized in Table I as compiled from the annual Extraction, transformation and loading (**ETL**) reports [1–5]. In 2018, cryptojacking was introduced to the list, and exploit kits were dropped [1]. The latter was dropping constantly in rank each year since it was at the fourth place in 2012 and 2013. Some of the top cyberthreats belong to same distinct threat category. For example, ransomware and cryptojacking are a specialization of the threat-type malware. Likewise, identity theft is a special category of data breach. Nonetheless, the overlapping threats are handled separately because the threat is launched by special malicious artifacts.

The top fifteen threats in 2017 have been the same top fifteen threats since 2014, although some order and trending have changed [1–6]. The top three threats have remained consistent since 2012, namely, malware, and web-based and web application attacks. Insider threat was not recognized as a distinct type of threat until 2013 while cyber espionage was not recognized until 2014. Position ranking was based on the number of incidents, impact, and relationship to other threats. A trend up indicates an increase in incidents but perhaps not an increase in rank. Table 11.1 shows the increased ranks of phishing and spam attacks as well as the decreased rank of exploit kit attacks which eventually exited the top fifteen threat list in 2018. The analyzed threat data are in the public domain, specifically from the Open Source Threat Intelligence Platform & Open Standards for Threat Information Sharing [7,8].

In addition to analyzing the type of threat, it is important to understand the source of the threat. The threats begin with humans and human motivation. The individual or organization must then gain access to the targeted computer or network. As pointed out in Refs. [1,6], threat sources consist of two entities:

11.3.1.1 Threat Agents

Threat agents are the actors, individuals, or organizations, who can create a threat. Often threat agents try to mask their identity and alliances by claiming identification with another group. This masquerading can be accomplished through fake news and social

---

TABLE 11.1

Annual Change in Ranking of the Top Fifteen Threats According to ETL

Top Threats	Year						
	2018	2017	2016	2015	2014	2013	2012
Malware	1	1	1	1	1	2	2
Web-based attacks	2	2	2	2	2	1	1
Web application attacks	3	3	3	3	3	3	3
Phishing	4	4	6	8	7	9	7
DoS	5	6	4	5	5	8	6
Spam	6	5	7	9	6	10	10
Botnets	7	8	5	4	4	5	5
Data breaches	8	11	12	11	9	12	8
Insider threat	9	9	9	7	11	14	–
Physical manipulation/damage/theft/loss	10	10	10	6	10	6	12
Information leakage	11	13	14	13	12	13	14
Identity theft	12	12	13	12	13	7	13
Cryptojacking	13	–	–	–	–	–	–
Ransomware	14	7	8	14	15	11	9
Cyber espionage	15	15	15	15	14	–	–
Exploit kits	–	14	11	10	8	4	4

media campaigns. In descending order, the most common categories of threat agents are cybercriminals, insiders, nation-states, hacktivists, cyberfighters, and terrorist groups.

11.3.1.2 Attack Vectors

Attack vectors are the path or means by which a threat agent gains access to a computer or network for the purpose of malicious activity. There is a large taxonomy of attack vectors. A short list includes the human element, web and browser attacks, Internet exposed threat, mobile app stores, and malicious USB drives.

11.4 TOP CYBERTHREATS

An analysis of the top threats shows an increase in the incidences of attack and attack tactics as well as advancements in defense. Ransomware attacks were a dominant threat. There was a massive increase in phishing attacks. In the sequel, we describe each of these cyberthreats in order of their ranking in 2018 and as described in Refs. [1,6].

11.4.1 THREAT 1: MALWARE

Malware is software with a malicious intent to destroy a computer, server, or network. It causes harm or acts against the interests of the user. Common malware are viruses, worms, Trojan horses, spyware, and ransomware. Malware remained the top cyberthreat since 2014. In 2017, over four million samples of malware are detected each day by malware protection firms [2]. An increase in malware attacks is escaladed by ad wars and hijacked browser sessions. Click-less infections, which do not depend on user interaction for deployment, are on the rise. The ETL describes 2017 as being highly mediatized with regard to leaked exploits. EternalBlue was one of several exploits attributed to the US National Security Agency (NSA) to be leaked by hackers. EternalBlue exploits a vulnerability in older versions of Microsoft Windows operating system and leads to a number of ransomware outbreaks. EternalBlue also serves as an example of the upward trending state intelligence development of malware. Noteworthy malware attacks of 2017 were WannaCry and NotPetya, both of which had a ransomware payload and exploited the EternalBlue vulnerability. The WannaCry ransomware attack lasted only a few days but is estimated to have resulted in hundreds of millions of dollars in damage in the USA, Japan, and Australia. WannaCry was attributed to North Korea. Malware's roles in the kill chain are installation of the threat, command and control of the device, and execution of harm. Mitigation of malware includes malware detection on all in-bound and outbound channels and sufficient security policies for response. In 2018, the attack vectors for detected malware were 92% by email compromise and 6% by web and browser.

11.4.2 THREAT 2: WEB-BASED ATTACKS

Web-based attacks make use of web-enabled systems such as browsers, webpages, and content managers. They are the most common threat for financial attacks and remained the second top cyberthreat since 2014. As a widely used content manager, WordPress is particularly vulnerable.

Drive-by download attacks involve malicious JavaScript and do not require action from the user. Malicious URLs use Blackhat Search Engine Optimization (SEO) to attract targets. A web-based attack's roles in the kill chain are the creation, delivery, and execution of a payload targeted to a particular vulnerability. Mitigation includes patching vulnerabilities and web traffic filtering.

11.4.3 THREAT 3: WEB APPLICATION ATTACKS

Web application attacks take advantage of application programming interfaces (APIs) which are exposed and open. Government and financial institution apps are particularly popular targets. SQL injection can be used to retrieve passwords stored in databases. Web application attacks remained the third top cyberthreat since 2012. Kill chain phases for web application attacks are reconnaissance for choosing targets and identifying vulnerabilities, exploitation of the vulnerabilities, and installation of access points which lead to command and control of the device. Mitigation includes policies for secure app development and for the authentication and validation of mechanisms.

11.4.4 THREAT 4: PHISHING

Phishing uses social engineering to lure targets into revealing-sensitive information. Spearfishing targets people within a specific organization. Often disguised as legitimate organizations, one million new phishing websites are created each month. Phishing's roles in the kill chain are reconnaissance, weaponization, and delivery. The human element is the weak link in the phishing attack. In fact, in 2018 over 90% of malware infections and 72% of data breaches in organizations originate from phishing attacks [1]. Mitigation includes the education of potential targets about fake email, random clicking, and oversharing of personal information.

11.4.5 THREAT 5: DENIAL OF SERVICE ATTACKS

DoS attacks occur when machine or network resources are made unavailable to their intended users by disrupting service, usually by flooding the network with requests, often from botnets. It is particularly damaging to organizations that rely on a web presence. Distributed denial of service (DDoS) attacks strike a target from many sources and are harder to stop. Pulse wave DDoS attacks come in short bursts on multiple targets and can last for days. The DoS attacks can be used to mask other attacks. In 2017, the Mirai Internet of Things (IoT) botnet was responsible for the largest DoS attack in history. The attack lent credence to warnings about IoT vulnerabilities and led to massive increases in security of IoT devices. These new measures led to some decrease in botnet activity and in DoS attacks, but the overall trend of DoS attacks is still increasing. Kill chain phases for a DoS attack are reconnaissance, weaponization, command and control of device, and execution of harm. Mitigation includes a reaction plan, Internet service providers (ISPs) with DoS protection, and organization specific protections such as firewalls and access control lists [9–12].

11.4.6 THREAT 6: SPAM

Spam is flooding users with unsolicited messages by email and messaging technologies. It comprised close to half of total email volume or nearly 300 billion to 450 billion emails per day in 2018 [1]. Most spam comes from botnets, 80% of which are thought to have been created by a group of about 100 spam gangs. In spite of the uncovering of several large botnets, spam attacks are still upwardly trending. Spam is the main means of malware delivery through attachments and URLs, although most spam is simply advertisement without malware. The roles of spam in the kill chain are weaponization and delivery. Mitigation is through spam filters and user education.

11.4.7 THREAT 7: BOTNETS

Botnets consist of several Internet connected devices, each device running a bot or script, that performs a simple task at high repetition. The IoT botnets were the second most important threat of 2017, responsible for an enormous DoS attack. Approximately 8.4 million new devices were added to the Internet in 2017. A large percentage connected devices are considered to be vulnerable. When such a device is compromised, it can become part of a botnet. There is a concern that virtual machines can become part of botnets. Fewer than expected DoS botnet attacks occurred in 2017, but the fear is that the focus of botnets has turned to ransomware. The kill chain phase of botnets is command and control of the device. Mitigation is through application and network firewalling and traffic filtering.

11.4.8 THREAT 8: DATA BREACH

Data breach is the loss of data that can only be discovered after the fact. It is not a threat but the result of a successful attack. Data breaches are likely more prevalent than known. A high number of breaches result from stolen or weak passwords and user credentials continue to be sold on the dark web. Breaches through espionage can be part of a nation-state cyberattack; however, 61% of data breach victims are small companies and 35.4% are in the healthcare sector. The EU's General Data Protection Regulation (GDPR) is intended to have an impact on careless breaches through repercussions. In the USA, Equifax's breach of nearly 150 million customers' personal and financial records in 2017 resulted in lawsuits and government investigation. In addition to penalties, data breaches are avoided by preventing cyberthreats. Mitigation includes encryption and reduction of access rights.

11.4.9 THREAT 9: INSIDER THREAT

Insider threat is the harm caused by an organization's insider with authorized access. The harm can be unintentional. This threat is thought to be on the rise, but losses are hard to quantify so the threat is deprioritized. Breaches are most commonly caused by high-level managers and outside contractors rather than non-managerial employees. In fact in 2018, 77% of the companies' data breaches are caused by insiders. Healthcare is a particular target with 59% of breached records resulting from insider threat. Insider threat involves all positions on the kill chain. User awareness is the most useful mitigation, in addition to segregation of duties and limiting access to data.

11.4.10   THREAT 10: PHYSICAL DAMAGE AND LOSS

Physical damage and loss of computers and equipment can result in data breaches. Encryption would solve the data breach problem, but only 43% of reporting organizations use encryption in 2018. Threat is more prevalent with IoT and mobile devices as device losses count for around half of all breaches in 2018. Additional examples of physical threat include automatic teller machine (ATM) drilling in which a drilled hole near the PIN pad allows for wired access and control of the machine. Theft of devices for their valuable components used to be limited to copper wire but now includes backup batteries in cell towers. In addition to encryption, mitigation includes asset inventory.

11.4.11   THREAT 11: INFORMATION LEAKAGE

Information leakage is a breach caused not by a direct attack but from unsecured data. Mobile devices make such breaches easier. Information leakage is usually the result of human error, often an insider action or failure. However, information can also be leaked through coding errors, particularly on mobile devices. In fact, unintended disclosure is the profound reason for information leakage in 2018, and human error is the most crucial factor for data disclosure [1].

In 2017, a navigation app installed on many Android devices contained a coding error that allowed hackers to obtain hardcoded credentials for text messaging. In 2018, information collected by the mobile fitness tracking and sharing app Strava has highlighted the locations of secret US military bases worldwide.

11.4.12   THREAT 12: IDENTITY THEFT

Identity theft is a cyberthreat in which the attacker obtains information about a person or computer system for the purpose of impersonating the target. Like a data breach, identity theft is the result of a successful attack. In the UK, identities were stolen at the rate of 500 per day in 2017. Personal and credit card data sell for as little as $10 on the black market. Yet most individuals report that they have few worries about identity theft. Top threats include skimmers on credit card devices, dumpster diving for hard copy personal information, phishing, hacking, and telephone impersonators. Mitigation includes protection of documents, strong privacy settings on social media, password protection on devices, and care when using public WiFi.

11.4.13   THREAT 13: CRYPTOJACKING

Cryptojacking (also known as cryptomining) is a new term that refers to the programs that use the victim's device processing power to mine cryptocurrencies such as Bitcoin, without the victim's consent. It is a type of malware, and unlike ransomware, the attacker is more focused on assuming the control of the machine's computational power and producing currency units indefinitely than being paid a ransom amount once. Cryptojacking made it to the top fifteen cyberthreat list in 2018. The number of victims was around half million per month in 2018.

11.4.14 THREAT 14: RANSOMWARE

Ransomware is a malware that encrypts files or locks down a system until the target pays the actor to remove the restrictions. In the case of wipeware, the encryption is never removed. Twenty percent of organizations reported that even after paying the ransom, they did not get back their data. Ransomware as a Service (RaaS) allows cybercriminals easy entry to ransomware attacks by providing a complete set of launch tools for less than $400 on the dark web. Ransomware ranking in the ETL top 15 cyberthreat list decreased significantly from 7th place in 2017 to 14th place in 2018 due to the shifting of focus by attackers to cryptojacking. In the latter, a computer is invaded in a way similar to ransomware, but instead of demanding a ransom, a malicious software is installed to start cryptocurrency mining without the computer owner's noticing. However, over 85% of the malware targeting medical devices in 2018 was ransomware [1]. Noteworthy ransomware attacks of 2017 were Cerber and Jaff. Cerber is a family of ransomware payloads which are distributed through email, exploit kits, JavaScript, and Microsoft Word macros. Variants of Cerber include a Bitcoin wallet stealing function. Jaff is described as new but vicious. It is spread via the Necurs botnet and is downloaded in a.pdf file attachment. Jaff is said to check for the target computer's language setting. If the language is set to Russian, the payload destructs rather than deploying. The kill chain roles of ransomware are installation, command and control of the device, and execution of harm. Mitigation is through limited access rights to data which potentially makes fewer data vulnerable to encryption, and an offline backup to recover data as well as an up-to-date and patched software and operating system [13].

11.4.15 THREAT 15: CYBER ESPIONAGE

Cyber espionage involves the use of a computer network to obtain confidential information. Government, political, and commercial organizations are the typical targets. Actors include nation-states and organized crime. Cyber espionage made it to the top 15 list in 2014 but remains at the bottom of the list. APTs are a collection of processes, tools, and resources used to infiltrate networks over a long period without detection. Even though cyber espionage is last on the top fifteen list, it is perceived as a serious threat, likely due to press coverage. Widely covered attacks in 2017 include the US Democratic National Committee breach, and the Ukraine power grid take-down attributed to Russian cyberspies. Since espionage is a composite threat, mitigation is through mitigation of all other cyberthreats [14,15].

11.4.16 THREAT 16: EXPLOIT KITS

Exploit kits are a bundle of ready-made exploits used to infect websites or as part of a malicious advertising campaign. The kits identify vulnerabilities on web browsers and web apps and then exploit automatically, an example of click-less attacks.

The payload may be ransomware. Common targets are Java and Adobe Flash add-ons. Exploit kits were in fourth place in the top fifteen threats in 2012, and increased in rank constantly until it exited the list in 2018. The scaling up of an exploit kit attack can lead to its detection which perhaps explains the trend. Mitigation is the detection and patching of vulnerabilities [16].

11.5 CONCLUDING REMARKS

Cyberthreats are constantly evolving and one cyberoperation may see multiple avenues of threats taken to fulfill objectives of the cyberactors. There is a desire for the implementation of security during development rather than after, particularly for IoT devices. That is "built-in" versus "bolt-on" security. The complexity and maturity of malicious practices will continue to be analyzed. There is a prediction that data analytics will be used not just to mitigate threats but to develop attacks. There is concern that state-sponsored and military cyberweapons will be tested on easy targets already in crisis through poverty or war. There is also a call for an increase in cyberthreat intelligence capabilities and training which is currently limited and lagging behind threats. In conclusion, a good reflection on the cyberthreats from an individual's standpoint may help educate ordinary users on prevention techniques to protect themselves, their organization, and their societies.

REFERENCES

1. Qadri, Y., Ali, R., Musaddiq, A., Al-Turjman, F., Kim, D., & Kim, S. (2020). "The limitations in the state-of-the-art counter-measures against the security threats in H-IoT". *Cluster Computing,*. doi: 10.1007/s10586-019-03036-7.
2. Jabbar, S., Khalid, S., Latif, M., Al-Turjman, F., & Mostarda, L. (2019). "Cyber security threats detection in Internet of Things using deep learning approach". *IEEE Access*, vol. 7, no. 1, pp. 124379–124389.
3. Wang, J., Jabbar, S., Al-Turjman, F., & Alazab, M. (2019). "Source code authorship attribution using hybrid approach of program dependence graph and deep learning model". *IEEE Access*, vol. 7, no. 1, pp. 141987–141999.
4. Gritzalis, D., Kandias, M., Stavrou, V., & Mitrou, L. (2014). "History of information: the case of privacy and security in social media." In *Proceedings of the History of Information Conference, Kuala Lumpur,* (pp. 283–310).
5. Deliri, S., & Albanese, M. (2015). "Security and privacy issues in social networks". In *Data Management in Pervasive Systems*, F. Colace et al. (eds.) (pp. 195–209). Springer, Cham.
6. Al-Turjman, F. (2020). "Intelligence and security in big 5G-oriented IoNT: an overview". *Future Generation Computer Systems*, vol. 102, no. 1, pp. 357–368.
7. Krishnamurthy, B. (2013). "Privacy and online social networks: can colorless green ideas sleep furiously?" *IEEE Security & Privacy*, vol. 11, no. 3, pp. 14–20.
8. Othman, N. F., Ahmad, R., & Yusoff, M. (2013). "Information security and privacy awareness in online social networks among UTeM undergraduate students". *Journal of Human Capital Development (JHCD)*, vol. 6, no. 1, pp. 101–110.
9. Carminati, B., & Ferrari, E. (2008, July). "Privacy-aware collaborative access control in web-based social networks". In *IFIP Annual Conference on Data and Applications Security and Privacy* (pp. 81–96). Springer, Berlin, Heidelberg.

10. Ali, S., Islam, N., Rauf, A., Din, I. U., Guizani, M., & Rodrigues, J. J. (2018). "Privacy and security issues in online social networks". *Future Internet*, vol. 10, no. 12, p. 114.

11. Ulusar, U. D., Al-Turjman, F., & Celik, G. (2017, October). "An overview of Internet of things and wireless communications". In *2017 International Conference on Computer Science and Engineering (UBMK)* (pp. 506–509). IEEE, Antalya.

12. Kauthamy, K., Ashrafi, N., & Kuilboer, J. P. (2017, April). "Mobile devices and cyber security – An exploratory study on user's response to cyber security challenges". In *International Conference on Web Information Systems and Technologies* (Vol. 2, pp. 306–311). SCITEPRESS, Porto.

13. Wang, X., & Yi, P. (2011). "Security framework for wireless communications in smart distribution grid." *IEEE Transactions on Smart Grid*, vol. 2, no. 4, pp. 809–818.

14. Vrhovec, S. L. (2016, May). "Challenges of mobile device use in healthcare". In *2016 39th International Convention on Information and Communication Technology, Electronics and Microelectronics (MIPRO)* (pp. 1393–1396). IEEE, Opatija.

15. Alabady, S. A., Al-Turjman, F., & Din, S. (2018). "A novel security model for cooperative virtual networks in the IoT era." *International Journal of Parallel Programming*, vol. 48, pp. 280–295.

16. Al-Turjman, F., & Zahmatkesh, H. (2019). "An overview of security and privacy in smart cities' IoT communications". *Transactions on Emerging Telecommunications Technologies*. doi: 10.1002/ett.3677.

12 Security in Grid and IoT-Enabled Cities

Fadi Al-Turjman
Near East University

Hadi Zahmatkesh
Middle East Technical University

Ramiz Shahroze
Near East University

CONTENTS

12.1 INTRODUCTION

In recent decades, the population of urban areas is rapidly increasing. Based on a report from the United Nations (UNs) Population Fund, more than 50% of population in the world inhabit urban environments [1]. The concept of "smart city" [2] has attracted too much attention by both academia and industry due to its strong requirements and practical background in an urbanized environment. Several cities have begun to develop their own strategies towards the concept of smart cities in order to enhance the quality of life and provide better services to citizens.

Many countries with growing population are spending a vast amount of money on smart cities-related projects. For example, China is working on more than 200 projects towards smart cities paradigm [3]. Smart cities-related technologies are enabling the urban municipals to manage their everyday operations to make people's life easier. Smart cities' infrastructure includes many devices and interconnected systems to benefit people in a variety of applications such as smart healthcare, smart transportation, smart parking, smart traffic system, smart agriculture, and smart homes to just name a few.

Information-centric networking (ICN) is a networking paradigm which is able to maintain packet delivery in unreliable environments. Therefore, ICN can be considered as an alternative for the IP-based networks in smart cities [4].

The integration of various low-cost smart devices such as sensors and actuators, and the rapid development of wireless communication technologies enabling small and low-cost objects to connect to the Internet have resulted in the rise in the deployment of IoT where physical objects are changing to smart objects in everyday life. Besides IP-based approaches such as the one presented in Ref. [5], ICN solutions can be applied to develop the emergence of IoT and its related applications. ICN is characterized as a concept in order to name content and locate information at the center of the architecture [6] rather than depending on the IP host identifiers. The principal idea is to completely change the Internet to a more generic and simpler architecture [7]. ICN can support various IoT scenarios and overcome their current limitations by utilizing its advantages in order to deploy various applications in heterogeneous environments such as smart homes and smart cities [8]. It can also be used as a framework to connect different objects with sensing capabilities to provide multiple services in the IoT environments. Moreover, the use of ICN can reduce the energy consumption in the IoT era [9].

Cities are being smart, and this may cause people to face huge security and privacy risks [10]. This is because of the nature of resource-constrained devices which makes the smart city vulnerable to different security attacks [11]. These vulnerabilities may cause several cyber-attacks in smart cities. For instance, malicious attackers may produce false data during the manipulation of sensing data which results in the loss of control over the highly intelligent systems [10]. In 2015, 230 thousand people in Ukraine suffered a major breakdown of electricity due to the attack of hackers to the smart grid which happened to the people in the form of denial of service (DoS) attack [1]. Many resource-constrained devices such sensors and cameras which are collecting and sharing sensitive data in smart cities can also be vulnerable to attacks by the malicious hackers threatening the security and privacy of people in smart cities. Due to these cyber-attacks, Home Area Information (HAI) that is collected and controlled through smart homes can provide a way to reveal people's lifestyle in terms of privacy and even result in economic loss [10].

According to a report, the market of smart cities is expected to gradually increase to $1.5 trillion by 2020 [12]. Actually, governments are responsible to attract large investments in order to fulfill the vision of smart cities [13]. This huge improvement includes deployment of thousands of sensor nodes in the city to provide real-time information to people about different services such as public transportation, traffic flows, the quality of water and air, and the energy consumption rate, to just name a few [14]. However, processing and analyzing the vast amount of sensitive data generate a number of security and privacy challenges and concerns regarding how to protect the sensitive data in the presence of unauthorized parties [10,15].

In the IoT era and smart cities, cloud computing can provide cost-effective services for data processing and storage. However, there are some issues in cloud-based IoT applications such as lack of mobility support, location awareness, latency, and security which can be resolved by fog computing paradigm [16,17]. Fog computing addresses these challenges by providing computing services to the users at the edge of the network, which in turn reduces latency and enhances the quality of service (QoS) [18]. However, security and privacy are challenging issues in fog computing due to the differences in fog computing and cloud computing which make the security solutions for cloud services not suitable for fog computing services available to the users. Various cryptographic techniques can deal with security attacks. However, these techniques are not appropriate for resource-constraint IoT devices in smart cities. One solution in this regard can be offloading additional operations related to security to a fog-based node, which can enable security and data analysis directly at the edge of the network [19]. In addition, in publishing and subscribing systems, which spread data from publisher to subscriber, publication is disseminated using a set of brokers, which can collect sensitive information of the users. In this regard, the system must ensure confidentiality of the publications and subscriptions while brokers are trying to have access to publications' tags and interests of subscribers [20].

According to a definition by IBM, the concept of smart city is based on three main characteristics called "instrumented," "interconnected," and "intelligent" [21], which are shown in Figure 12.1.

Instrumented: This characteristic means a city which is covered with a group of devices such as sensors, and actuators. Therefore, the core systems of the cities have access to reliable and real-time information using these devices.
Interconnected: It means that smart city has a huge set of systems that are cooperating to provide information from various locations and sources. It is then possible to create a link from the physical world to the real world by using an accurate combination of interconnected and instrumented systems.
Intelligent: It refers to an instrumented and interconnected environment that utilizes the information obtained from various systems and devices such as sensors to improve the quality of life of the citizens.

12.1.1 COMPARISON OF SIMILAR SURVEYS

In spite of the aforementioned benefits of smart cities, several security and privacy concerns are arising due to the large number of wirelessly connected sensors and cameras which are collecting and sending data to base stations (BSs) and other Internet devices used to process data. All these devices are generating sensitive data over different networks. Data is the most precious asset of people in today's smart world. All data is handled by software and hardware that have some security and privacy issues such as the vulnerabilities in infrastructures and cyber-attacks (e.g., DoS attacks). Due to these security issues, the performance of highly innovative systems can degrade in the form of services. It is important to overcome these

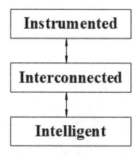

FIGURE 12.1   Main characteristics of smart cities.

security challenges to make the future of the highly advanced system more secure and beneficial for the users [22].

There are several published survey papers related to the security and privacy of IoT and smart cities. For example in Ref. [10], security and privacy in smart cities' promising applications are investigated. The authors also discussed several security and privacy challenges in these applications. In Ref. [23], the authors identified the techniques used for data security and privacy and discussed the technologies that make smart cities a reality. The study in Ref. [24] discussed different privacy types, attackers, and the required sources for the attacks in smart cities. The authors also reviewed the current privacy-improving techniques as well as various types of citizens' privacy in smart cities. Similarly, the main research challenges and the current security solutions in the IoT environments are presented in Ref. [25].

Our contributions in this study relative to the recent literature can be summarized as follows:

- This chapter highlights main applications of smart cities and addresses the major privacy and security issues in the architecture of the smart cities' applications.
- We provide a number of solutions in order to deal with critical threats regarding the security and privacy in smart environments.
- We also propose a secure IoT architecture for the smart cities.
- Security and privacy techniques dealing with the process of developing secure systems are also discussed in the manuscript.
- Finally, we provide some open research issues and challenges that should be taken into account regarding the improvements of the smart cities in terms of security and privacy.

The remainder of this chapter is organized as follows. Section 12.2 presents an IoT-based architecture that focuses on the security and privacy issues in smart cities. Some typical applications of smart cities are outlined in Section 12.3. Sections 12.4 and 12.5 discuss a number of security and privacy issues and solutions in smart cities' environment, respectively. The general requirements regarding the security and privacy challenges for smart cities' services are presented in Section 12.6. Section 12.7 proposes a secure IoT-based architecture for smart cities. Section 12.8 discusses some open research issues and provides future research directions. Finally, Section 12.9 concludes this chapter. A list of abbreviations together with their brief definitions used throughout the chapter is provided in Table 12.1 to help the readers in understanding the abbreviated terms.

12.2 ARCHITECTURE OF SMART CITIES

In this section, we provide an IoT-based architecture that emphasizes on the security and privacy issues in smart cities. This architecture is built upon the architecture proposed in Ref. [26] and is shown in Figure 12.2. A brief discussion of each layer of the architecture is provided in the following subsections.

TABLE 12.1
List of Abbreviations

Abbreviated	Name
6LoWPAN	IPv6 over low-power wireless personal area networks
AA	Availability attack
AI	Artificial intelligence
AS	Autonomous system
AV	Autonomous vehicle
BLE	Bluetooth Low Energy
BN	Black Network
BS	Base station
CA	Content analysis
CCTV	Closed-circuit television
CoAP	Constrained Application Protocol
CS	Crowed sensing
DDoS	Distributed denial of service
DoS	Denial of service
EVC	Electronic vehicle charging
FI	False information
GPS	Global Positioning System
HAI	Home Area Information
HE	Homomorphic encryption
HetNet	Heterogeneous network
HIS	Intelligent Healthcare System
HTTP	Hypertext Transfer Protocol
ICN	Information-centric network
ICT	Information and Communication Technology
IDS	Intrusion detection system
IoT	Internet of Things
IS	Identity Spoofing
ITS	Intelligent transport system
KMS	Key management system
LTE	Long-Term Evolution
MCS	Mobile crowed sensing
ML	Machine learning
M2M	Machine to machine
MM	Message modification
MQTT	Message Queuing Telemetry Transport
P2P	Peer to peer
PPBS	Privacy-preserving biometric scheme
PPDM	Privacy-preserving data mining
QoS	Quality of service
RFID	Radio-frequency identification
SDN	Software-defined networking
SG	Smart grid
SSL	Secure socket layer

(Continued)

TABLE 12.1 (*Continued*)
List of Abbreviations

Abbreviated	Name
TA	Traffic analysis
TTP	Trusted third party
UAV	Unmanned aerial vehicle
UDP	User Datagram Protocol
UN	United Nations
UR	Unified Registry
VoIP	Voice over IP
WSN	Wireless sensor network

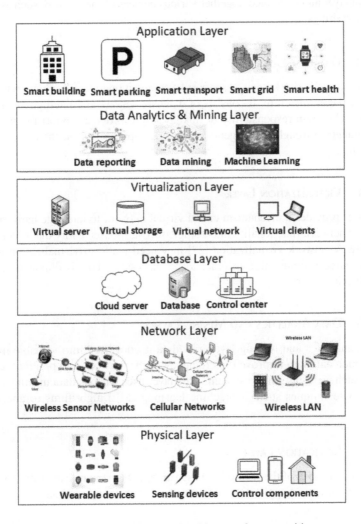

FIGURE 12.2 An overview of the IoT-based architecture for smart cities.

12.2.1 PHYSICAL LAYER

Physical layer is also known as perception layer or lower layer of the architecture. This layer contains heterogeneous devices (e.g., sensors and actuators) that collect information and send it to the upper layer of the architecture called network layer for further processing.

12.2.2 NETWORK LAYER

Network layer is also known as communication layer which is the core layer of the IoT-based architecture. This layer is dependent on basic networks such as wireless sensor networks (WSNs) and the Internet, as well as communication networks. The main responsibility of the network layer is to transmit the collected data by the physical layer and to connect together various devices of the network such as servers and smart things.

12.2.3 DATABASE LAYER

Database layer is also known as support layer, and it operates closely with the upper layers of the architecture. It consists of database servers and intelligent computing systems. The main responsibility of this layer is to provide support for application requirements through intelligent computing approaches such as cloud/edge computing.

12.2.4 VIRTUALIZATION LAYER

This layer provides a mechanism called virtual network to integrate hardware/software and network functionality into a single software-based entity that is configured logically. A network virtualization [27] may need platform virtualization together with resource virtualization in order to be successful. This is obtained using the virtualization layer.

12.2.5 DATA ANALYTICS AND MINING LAYER

In data analytics and mining layer, raw data is converted into valuable information, which can help improve the efficiency of the network and predict the future events such as failure in the system. This layer uses various data mining and data analytics techniques such as machine learning (ML) algorithms to analyze the data.

12.2.6 APPLICATION LAYER

This is the top layer of the secure IoT-based architecture and is responsible to provide intelligent and smart applications and services to the users based on their individual requirements. Some typical ones of these applications are briefly described in the following section.

12.3 APPLICATIONS OF SMART CITIES

Various applications of smart cities have been emerged to monitor the physical world. These applications are able to sense and collect information through thousands of smart devices (e.g., sensors) via Internet to make people's lives much easier and effective in different aspects such as environment, energy, transportation, safety, healthcare, parking, and traffic systems. In the following subsections, we describe some typical ones on these applications.

12.3.1 SMART GRID

Smart grid (SG) is the next generation of the electric grid and has been often used to refer to applications of power grid such as the peer-to-peer (P2P) energy trading [28]. SG is a data communication network, which provides a smart approach to integrate conventional power generation, energy storage, transmission, and distribution, as well as demand management to enhance reliability and provide higher operational efficiency and better power quality [29]. SG is an intelligent monitoring system that controls the electricity flowing via the grid system. This uses devices that have the capability of bi-directional communication to measure and sense production and consumption of electricity. It then sends the information to the operators and users as well as automated devices for monitoring and making decision regarding any changes in the condition of the electricity grid [30].

12.3.2 SMART TRANSPORTATION

The aim of smart or intelligent transportation is to provide a smarter use of transport system such as the infrastructure for electronic vehicle charging (EVC) [31,32]. Smart transportation consists of intelligent networks that can serve people by improving safety, reliability, and speed [33]. By using smart applications such as transport-oriented smart phones, people can easily search the fastest and the most economic routes, schedule their visits, and easily find the location of buses and trains. Smart transportation also facilitates car-parking searching and license recognition systems [34].

12.3.3 SMART ENVIRONMENT

Smart environment can have significant contribution towards the building of sustainable societies for smart cities. By using technical management devices, smart cities are able to monitor air quality, traffic congestion, and energy consumption, and to enhance waste and pollution efficiency [35]. Moreover, smart environment can monitor greenhouse gases, forest condition, city noise, etc. to entail the sustainable and intelligent development. It may also be possible to forecast and discover disasters in the future by using environmental WSNs [36].

12.3.4 SMART LIVING

Smart living provides intelligent management of different home appliances in order to enhance energy efficiency and provide convenient home [37]. It can also support remote monitoring of home appliances, energy savings, education, and entertainment. In addition, the applications of smart living can also manage the process waste recycling, and parking to provide a smart building with convenient life and great experience as well as sustainable energy and environments for the residents [10].

12.3.5 SMART HEALTH

In smart cities, the concept of smart health is to provide the health services by using networks and sensing infrastructures of smart cities [38]. Intelligent healthcare system (IHS) provides health monitoring and proper diagnosis to the people in smart cities [39]. The health conditions of the people can be timely monitored by using medical sensors and wearable devices [40]. The health data can then be forwarded to the processing unit for further diagnosis of the doctors. Moreover, the complete health-related information of the patients can easily be accessed through a database which in turn increases the possibility of diagnosing infectious or chronic illnesses in the early stage [10].

12.2.6 SMART ENERGY

In smart cities, sensor nodes are widely deployed to monitor energy generation and consumption. In this regard, smart energy [41] leveraging smart grid and EVC can reduce the energy consumption and stop the failure of the electricity supply in power grid as well as individual energy usage [10].

12.4 SECURITY AND PRIVACY ISSUES IN SMART CITIES' APPLICATIONS

Recently, significant problems have been occurred in various application scenarios. For instance, in smart grids, the smart metering infrastructure can control the private lives of the citizens such as their working hours in smart cities [42]. Furthermore, in smart homes and healthcare context, service providers and device manufacturers can have access to sensitive information of the users [43]. Moreover, smart mobility applications can collect huge amount of information related to trajectory of a user, which can be used to predict the mobility pattern and location of the user [44]. In addition to these problems, security and privacy have become a major challenge in smart cities-related applications since cities are focusing to become smarter. In recent decades, several security and privacy issues have been found in various smart cities-related applications [1]. In the following subsections, we list and briefly discuss some major ones of these security issues.

12.4.1 CYBER-SECURITY

Cyber-attacks compromise security of smart cities' applications and are mainly of two types: active attack and passive attack. The aim of passive attack is to learn and use different information of the system without any changes in resources of the system. The main target of this attack is "transmitted information" for the purpose of learning the configuration and behavior of the system and its architecture. It is hard to detect these attacks because the data is not modified. That is why it is better to focus more on the prevention of such attacks. On the other hand, the active attacks are scheduled to produce an effect or change in the operation of the system using data modification or adding incorrect data into the system. Sabotage, manipulation, and espionage are the main reasons behind cyber-attacks [45]. The main cyber-attacks, which may occur in various smart cities' application, are briefly described in this section as follows [46]:

Denial of Service (DoS): DoS can also be called "availability attack (AA)." The main purpose of DoS attack is to suspend the communication of the system. To do so, attacker can disable the physical component access through the excessive messages on the communication network which prevents the normal operation of the system. DoS attacks are a type of attacks, which can destroy the availability of the targeted system in smart cities. This type of attacks can be classified into network layer and application layer DoS attacks [47]. Network layer DoS attacks are performed at the network layer, and they try to overwhelm the resources of the network of the targeted system with bandwidth-consuming attacks such as User Datagram Protocol (UDP) flooding attacks. The application layer DoS attacks, on the other hand, utilize the special characteristics of the application layer protocols such as Hypertext Transfer Protocol (HTTP) and voice over IP (VoIP) in order to affect on the resources [48]. In smart cities, the impacts of both types of DoS attacks on any system that provides centralized monitoring in these areas can be extremely bad since the unavailability of the system would result in total chaos in the cities [49].

Malware: This is a malicious software that can gain illegal access to the system. It can also use internal weaknesses of the system aiming to steal, change, and ruin physical system components and related information. For example, smart cities may contain several closed-circuit television (CCTV) cameras controlled either privately or by public authorities. The security of these cameras is a challenging task as some of them lack encryption algorithms, and others are vulnerable to attack by malware [50]. Accessing a camera can provide a way to view individual's homes or use a bank camera to control and view the digits being pressed by the users.

Eavesdropping: This is an instance of a passive attack which is defined as an illegal listening to a communication without the permission of the communication's parties. Eavesdropping is a dangerous attack in smart cities that results in breaking down the confidentiality and integrity of the network and can lead to personal and financial failures [51]. It can be used to spy on communication channels in order to capture the behavior of the network traffic and obtain the network map.

Masquerading: It is also known as "impersonation" or "Identity Spoofing (IS)." In masquerading, the attacker tries to steal information by pretending to be a legal device or entity. For example, in intelligent transport system (ITS) as an application of smart cities, masquerading can provide unauthorized access to restricted information, which may ruin the integrity of the network. It may also result in loss, corruption, and manipulation of information in ITS [52].

False Information (FI): Attackers transmit erroneous amount of FI in the network, which might affect the behavior of other drivers. It can be both intentional and unintentional. Introducing FI on the systems in smart cities may result in delays and unnecessary congestions as people act based on the FI provided to them.

Message Modification (MM): In this attack, the message is modified to make an unexpected behavior happen in the system. MM may also include reordering a stream of message and/or message delay. Similar to FI, MM can cause unnecessary congestions and delays in the systems and threaten data integrity. As a result, threat to data integrity may harm people and infrastructures of smart cities [53].

Traffic Analysis (TA): TA is similar to eavesdropping, but here, rather than content analysis (CA), attackers monitor the traffic pattern and obtain useful information from it. Combining TA with eavesdropping may damage privacy. In addition, obtaining illegal information from TA attacks can harm confidentiality of information in smart cities.

Summary of the cyber-security attacks, which may occur in smart cities, is presented in Figure 12.3.

12.4.2 BOTNET ACTIVITIES IN IoT-BASED SMART CITIES

Botnets are one of the latest issues which have recently posed serious risks to IoT-based systems. An example of such botnets would be Mirai botnet which modifies or destroys information on various devices such as routers, webcams, IP cameras, and printers, and sends the infection to various IoT devices. This may finally result in a distributed denial of service (DDoS) attack against the target servers [54]. IoT devices are usually designed with almost no security compared to other smart devices such as smart phones and computers. This danger was not recognized until 2016; therefore, more research is required to overcome this threat in the future. Otherwise, this attack will damage the IoT ecosystem [55]. An approach to prevent such attacks is presented in Ref. [56]. The authors proposed an approach called BotDet for botnet command and control traffic detection in order to protect critical systems against malware attacks. They developed four detection modules to discover various techniques utilized in botnet command/control communications and designed a system for alert correlation

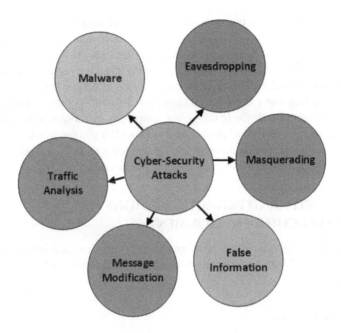

FIGURE 12.3 Summary of the cyber-security attacks in smart cities.

according to the voting between the detection modules. The results reveal the capability of real-time detection in the approach and show that the proposed system balances the false- and true- positive rate with more that 13% and 82%, respectively.

12.4.3 THREATS OF UNMANNED AUTONOMOUS VEHICLES IN SMART CITIES

Driverless cars or autonomous vehicles (AVs) are a type of car that drive itself without the human involvement using various in-vehicle technologies and devices such as sensors, Global Positioning System (GPS), and anti-lock braking system. AVs have gained too much attention towards the building of a smarter society with the goal of reducing the ratio of traffic accidents. Despite all the benefits of this growing application, there are major security and privacy risks once it is hacked. Because it will then threaten data privacy and life safety of the citizens in smart cities [57]. Particularly, hackers can use the security bugs to control the vehicle in order to shut down the engine or apply the brakes in various situations. Moreover, the huge amount of personal data obtained from the computer system of an AV may cause crucial security and privacy issues.

12.4.4 PRIVACY LEAKAGE

Attackers can collect, transmit, and process private information such as health conditions, identity and location of the users in transportation context, and lifestyle derived from the intelligent surveillance systems. This in turn makes the smart cities vulnerable to privacy leakage. To overcome this issue, a number of security and privacy solutions (e.g., anonymity, access control, and encryption) can be employed to save sensitive information in smart cities against the attacks of hackers [58,59]. However, most of the existing security and privacy techniques are designed only against outside attackers and do not consider the role of potential inside attackers [59]. For example, a smart building may use a surveillance camera to discover robbery or unusual activities. Attackers from inside the building (e.g., employees or those who have access the surveillance records) may steal private data or provide a gap for attackers outside the building. Therefore, it is a challenging task to develop a security and privacy mechanism in smart cities to have a balance between privacy and efficiency.

12.5 SECURITY AND PRIVACY SOLUTIONS FOR SMART CITIES' ENVIRONMENT

In this section, a number of security and privacy solutions are introduced which are utilized to deal with critical threats regarding the security and privacy in smart environments.

12.5.1 BLOCKCHAIN

Blockchain is a P2P distributed, centralized, and public decentralized technique that is used to store transactions, sales, agreements, and contracts across many computers [59]. Blockchain is basically a chain of blocks, where digital information is stored in a public database. It is specially developed for crypto-currency such as Bitcoin and Litecoin. Blockchain technology has gained more attraction in recent years. The comprehensive survey in Ref. [60] verifies the reliability of utilizing this technology in IoT environments and reveals its significance in the developing IoT ecosystems. The main reason behind the success and importance of blockchain technology in IoT applications is its decentralized feature which enables various applications to work in a distributed manner. For instance, in Ref. [59], the authors proposed a security framework based on the blockchain technology that can secure the communication of different devices in smart cities and enhance the effectiveness and reliability of the system. Moreover, a new security framework is developed by the integration of blockchain technology into a smart home in order to improve integrity, confidentiality, and availability [61]. Similarly, security and privacy issues in vehicular communications are addressed using blockchain technology in Ref. [62]. In Ref. [63], the authors proposed a personal data management system that integrates blockchain technology with off-blockchain storage solution so that the users are aware of the collected data by the providers. The authors in Ref. [64] proposed an auditing system based on blockchain for shared data in cloud-based applications. They presented an approach where a number of entities should collaborate to get back the identity of a malicious user. Based on the proposed blockchain-based architecture, data changes can be traced, and correct data blocks can be recovered when the data are damaged. In addition, various design requirements of solutions based on blockchain technology for data origin tracking are discussed in Ref. [65]. The authors presented the assessment of their implementation results in order to provide a complete overview of various specified approaches. In Ref. [66], a blockchain-based access control is presented that allows owners of data to specify access rights for the data sources on remote servers and change the privileges when required. This approach provides an authenticated access control managed by blockchain technology that ensures the preservation of users' privacy. Furthermore, blockchain technology can provide robust solutions and secure smart cities from cyber-attacks. For example, blockchain-based digital smart ID can be assigned to everyone and everything in order to provide authentication and authorization for people as well as connected devices within smart cities' network.

12.5.2 CRYPTOGRAPHY

Algorithms utilizing cryptography techniques are the backbone of the security and privacy protection in information-centric smart cities' applications since they prevent the access of unauthorized parties during the data storing, transmitting, and processing. The existing cryptographic tools utilized in smart cities' applications are discussed in this section. Traditional encryption algorithms and standards are not fully appropriate for resource-constrained smart devices due to the energy consumption and computational complexity [67]. Therefore, it is a basic requirement to use lightweight encryption for utilizing cryptographic algorithms in practice. For instance in Ref. [68], the authors proposed a mechanism for IoT-based scenarios that can secure end-to-end communications of the users from DDOS attacks. Moreover, a lightweight authentication protocol was developed in Ref. [69] to secure smart cities' applications by using a public key encryption strategy. In addition, homomorphic encryption (HE) has recently attracted too much attention due to its strength and capability on computations of encrypted data. For instance, HE can be utilized for the protection of electricity consumption in smart grid systems [70] as well as for solving security and privacy issues in cloud computing [71]. It can also be used to protect security and privacy in healthcare monitoring systems [72]. Moreover, cryptography approaches can be considered as one of the most convenient and effective techniques to provide security for cloud-based data as they significantly improve the security and privacy of data particularly in public cloud environment [73]. In addition, cryptographic techniques support well-known privacy preservation algorithms and are able to offer very precise analysis results [73]. However, it may also be possible to attack cryptographic algorithms with the help of simple power analysis. Attackers may control and change the data by intentionally injecting errors trying to have an effect on the performance of the device [74]. Therefore, it is important to carefully conduct research in cryptographic algorithms to ensure the security and privacy of the data. In addition, cryptography can be used to solve the problem of privacy leakage caused by public access policy [75]. For example, the authors in Ref. [76] proposed a scheme to consider a trade-off between decryption possibility and the policy privacy in which the attribute information in the policy is divided into two parts: value and name. In the proposed method, the value attribute is hided in the access policy instead of concealing the whole attribute. Therefore, it can protect the policy privacy reasonably.

12.5.3 Biometrics

Biometrics are broadly used for authentication in IoT-based infrastructures. This technology widely depends on human behavior and automatically recognizes a person through bio-data obtained from faces, fingerprints, handwritten signatures, voices, etc. One of the most accurate methods, which can obtain high accuracy and efficiency, is brainwave-based authentication [77]. Similarly, a mutual authentication protocol purposed in Ref. [78] to keep safe the private information of the users in storage devices. Please note that the risk of privacy leakage would increase if the mentioned bio-based approaches are not used properly. For example in Ref. [79], the authors reported that it is required to develop privacy-preserving biometric schemes (PPBSs) similar to the one presented in Ref. [80]. Moreover, they revealed that these biometrics have promising use cases in the future in various applications such as e-business. Biometrics can also be used to encrypt communication between an unmanned aerial vehicle (UAV) and BS. For example, the study in Ref. [81] proposed a safety mechanism with low-cost resources based on biometrics for the UAV if an attack is detected. The authors showed that the proposed approach could be applied to any UAV scenario where the cyber-security attacks are an important issue.

12.5.4 Machine Learning (ML) and Data Mining

ML is a part of Artificial Intelligence (AI) with the goal of developing systems which are able to learn from past experience. According to the current situations, ML techniques can be used to enhance the efficiency of intrusion detection systems (IDSs) in IoT environments [1]. For example, the study in Ref. [82] showed the advantages of using ML technologies to provide security in WSNs. Similarly, in Ref. [83], the authors developed a machine-based approach to improve security of data sensing in WSNs. In addition, a novel model utilizing ML algorithms is developed in Ref. [84] to discover attacks in Wi-Fi networks. There are also several studies that employed ML technologies to strengthen defense-related strategies. For instance, in Ref. [85], the authors developed a model based on game theory and ML technologies to discover and prevent intrusions in WSNs. Moreover, the existing studies on biometric security systems from the ML point of view are reviewed in Ref. [86]. In addition, there are different ML techniques such as supervised learning and unsupervised leaning that can be applied to effectively detect the presence of a botnet [87]. However, there are still some issues such as real-time monitoring and adaptability to new attacks that need to be solved regarding ML-based detection techniques.

Data mining is another technique that can be applied to handle security and privacy in smart city environments. The study in Ref. [88] showed that the huge amount of data collected by various sensors in smart cities are utilized to mine new information and regulations which in turn provides better services to the users. However, using data mining techniques may result in some security and privacy concerns regarding the disclosure of sensitive information of the users such as their locations. In this regard, privacy-preserving data mining (PPDM) techniques can be applied in order to overcome this problem [89,90].

TABLE 12.2

A Summary of the Security and Privacy Solutions in Smart Cities' Environments

Ref	Blockchain	Cryptography	Biometrics	ML
[10,24]	-	X	X	-
[59,61]	X	-	-	-
[68,69]	-	X	-	-
[79,80]	-	-	X	-
[23]	X	X	-	X
[82,83]	-	-	-	X
[60,62]	X	-	-	-
[25]	-	X	X	-
[84,85]	-	-	-	X

ML = Machine learning.
X = Considered.
- = Not considered.

12.5.5 IoT REGULATIONS

IoT includes a wide range of communication technologies such as machine-to-machine (M2M) communication, sensors, wireless communication, and radio-frequency identification (RFID). However, the IoT industry is still not regulated. This has resulted in broader security and privacy implications. The use of unsecured smart devices in various industries such as military and health, and the fact that IoT devices can be easily hacked have created new attacks which can happen at each layer of the IoT protocol stack [91]. While the IoT industry has been aware of the issues related to security and privacy, recent cyber-attacks on IoT devices such as the one presented in Ref. [92] and other similar attacks had unintended results of increasing the awareness [93] regarding the needs of having strong security mechanisms and regulations for the devices that are connected to the Internet.

A summary of the aforementioned solutions for smart cities' environments is presented in Table 12.2.

12.6 SECURITY AND PRIVACY REQUIREMENTS FOR SMART CITIES' SERVICES

In this section, we briefly discuss security and privacy techniques dealing with the process of developing secure systems.

12.6.1 PRIVACY BY DESIGN

This is a strategy trying to fix the security and privacy issues in information-centric smart cities [94]. This strategy includes some principles that should be taken into account while designing a new system [95]. For example, there should be a proactive privacy protection rather than a post-reaction after the violations happen. Moreover, privacy should be considered in the design of the system and available as the default setting. In addition, there should be protection for the whole lifecycle of the data. Finally, the system should have transparency and visibility, and respect the user privacy. Several studies utilized these principles in developing new privacy-friendly systems. For instance, the study in Ref. [96] utilized proactivity principle in the design of the solution for a remote health monitoring system. Moreover, the study in Ref. [97] applied the principle of visibility and transparency to ITSs.

12.6.2   TESTING AND VERIFICATION

This is a crucial part of the design of a security- and privacy-friendly system to make sure that the implementation of such systems achieves its security- and privacy-related requirements. Privacy-related testing and verification should be incorporated into the current testing processes as they are not basically different from other types of testing [24]. The main goal of these verification approaches is to find information leaks from different applications, for instance, by using black box differential testing [98]. The process of testing and verification must be applied in design of any new system architecture in information-centric smart cities.

12.6.3   PRIVACY ARCHITECTURE

Privacy architecture is required to consider various protection approaches to ensure that there are no privacy leakages in the system. For instance, the study in Ref. [99] proposed an architecture relying on the trustworthy remote data stores as well as a broker which intercedes access to the users' data stores. In addition, the study in Ref. [100] combined various cryptographic approaches to provide privacy in the system.

12.6.4   DATA MINIMIZATION

According to Ref. [101], data minimization should be considered as one of the most important parts of the privacy by design strategy. In information-centric smart cities, this strategy can be utilized in various applications such as electronic toll pricing system in order to analyze different architectural options [102]. It can also be used in the analysis of big data to obtain privacy-protection solutions [103]. In smart cities, smart systems should be designed in a way to avoid recoding unrelated data. For example, cameras for ITSs can also record unrelated information or sensors in smart environments may collect more data rather than needed. Therefore, data minimization techniques can be used to overcome such challenges.

12.6.5   SECRET SHARING

This method allows distribution of secret information among different participants [24]. It is usually divided into m shares with each participant having one share. This method requires at least n shares to recover the secret. Therefore, it provides reliability and confidentiality in the system. In information-centric smart cities, secret sharing can be utilized for distributed data storage [104], and for data aggregation from smart meters [105] and sensor networks [106].

12.6.6 SYSTEM SECURITY AND ACCESS CONTROL

In information-centric smart cities, security of the system and its sub-systems is crucial for the purpose of privacy protection. For examples, if there are vulnerabilities in the system, attackers can easily have access to smart devices and retrieve the desired data which would be very dangerous for the privacy of users' sensitive data. Therefore, it is mandatory to secure the system in order to avoid any attacks from the hackers. Access control restricts the access of data from unauthorized parties. It would also help to minimize control of the system from the misuse of stored data. Access control is also important for autonomous systems (ASs) with a connection to the Internet in which smart devices can be monitored remotely [107].

12.6.7 SECURE MULTI-PARTY COMPUTATION

It is a cryptographic approach that permits multiple parties to calculate the value of a public function, without showing the private inputs of the parties and without depending on a trusted third party [24]. In smart cities, this method can be used in the design of a healthcare system to analyze the results of genomic tests where it is required to keep the test sequence and the patient's genome private [108].

12.7 RECOMMENDED SECURE ARCHITECTURE

An IoT-based smart city architecture operates over heterogeneous networks (HetNets) and consists of millions of resource-constrained devices. In Figure 12.4, basic components of an IoT-based smart city architecture are shown. These components include Black Networks (BNs), trusted software-defined networking (SDN) controller named as trusted third party (TTP), Unified Registry (UR), and Key Management System (KMS) [109]. These four components are responsible for secure communication and authentication across HetNets and have different responsibilities in the architecture. BNs are responsible for data privacy, integrity, confidentiality, and authentication. TTP is responsible for efficient routing across IoT nodes, whereas UR is used for a database of various devices such as nodes, sensors, and gateways. Finally, KMS is responsible for IoT networks.

12.7.1 BLACK NETWORKS (BNs)

BNs are used to secure data that contains the meta-data, related to each packet in an IoT protocol [110]. They can secure data through various encryption methods which can be done viaGrain128a or AES in the EAX or OFB modes. BNs enable authentication and secure communication at both link layer and network layer [109]. Moreover, BNs can reduce a wide range of active and passive attacks which in turn provides confidentiality, integrity, and privacy in IoT-based networks because of the secured communications at the network and the link layer.

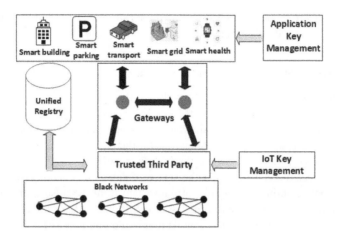

FIGURE 12.4   An overview of the components of a secure IoT-based smart cities architecture.

12.7.2 TRUSTED SDN CONTROLLER

SDN is a paradigm that provides several opportunities to secure the network more efficiently. In this architecture, a SDN controller is used for the communication of network
devices using various protocols. OpenFlow is the most commonly used protocol for the
communication between various devices in the network and the SDN controller [111].
The main aim of the SDN controller is to resolve the routing challenges in privacy protection of IoT-based BNs [110]. OpenFlow protocol allows the SDN controller to begin
a secure connection with the devices in the network. Trusted SDN controller can manage sleep and/or wake cycles, and can keep up a global IoT-based network view [109].

12.7.3 UNIFIED REGISTRY (UR)

The main purpose of UR is to consolidate the heterogeneous technologies to create
IoT-based networks for smart cities. The concept can also be extended to a visiting UR for the IoT nodes that are mobile and cross-systems. This is crucial from
a security perspective that a variety of IoT networks consider fixed nodes communicating through wireless communication technologies. There are several wireless
communication technologies utilized in smart cities including W-Fi and Long-Term
Evolution (LTE). Moreover, many protocols such as IPv6 over low-power wireless
personal area networks (6LoWPAN), Zigbee, and Bluetooth Low Energy (BLE), and
several addressing approaches such as IPv6 128-bit addressing, RFID addressing,
and Bluetooth 48-bit addressing can be utilized in smart environments. All these
technologies, protocols, and schemes require a unified attribute for identity management, authentication, and authorization. Furthermore, translation between wireless
communication technologies, protocols, and addressing approaches has to be done,
and UR makes the conversion process easier. It is difficult to implement a UR due
to several regulatory, practical, and security issues. However, a logical entity can be
implemented in a highly distributed manner that focuses on data and attribute set of
an IoT node within the smart cities' network.

TABLE 12.3

Components and Services of a Secure IoT-Based Architecture for Smart Cities

Components	Services
BNs	Integrity, confidentiality, privacy
TTP	Availability, secure routing
UR	Identity management, mobility, authentication, authorization
KMS	Efficient key distribution for symmetric keys

12.7.4 KEY MANAGEMENT SYSTEM (KMS)

KMS [109] is the process of managing different tasks associated with various aspects of cryptographic key for cryptosystem. It is an important part of all security infrastructures. In an IoT environment, resource-constrained devices communicate in a secure manner using a symmetric shared key. Keys should be created and stored, and used securely. A critical issue regarding symmetric keys in a distributed mobile system is related to key distribution. Using a hierarchal KMS provides efficient key distribution for symmetric keys in smart cities. It is required to authenticate communications between IoT devices [112]. Moreover, IoT applications require different security protocols and standards as they may deal with different security vulnerabilities. Some of the most commonly used IoT protocols such as Message Queuing Telemetry Transport (MQTT) and Constrained Application Protocol (CoAP) do not have built-in security mechanism [113]. However, CoAP and other protocols can utilize protocols such as secure socket layer (SSL) to enhance their security. This means that for secure communication between IoT devices, it is not sufficient to have single factor authentication [113]. Therefore, it is required to have the KMS with two-factor authentication mechanism in order to mitigate the risk and improve the security in IoT networks [114].

As an example for such an architecture, consider a node in an IoT BN that wants to send a packet to another node. The SDN controllers can ensure availability and synchronize the nodes for routing by creating the flow tables for the packets to be routed from one node to another. In this regard, UR is responsible for identity management as well as node authentication and authorization. In addition, KMS provides external key management for the IoT nodes in order to communicate securely using a shared key.

A summary of the security components and services in IoT-based smart cities is provided in Table 12.3.

12.8 OPEN ISSUES AND FUTURE RESEARCH DIRECTIONS

In spite of different research studies and rapid improvement that have been obtained in recent decade regarding the security and privacy in information-centric smart cities, there are still open research issues and challenges that should be taken into account regarding the improvements of the smart cities in terms of security and privacy.

12.8.1 Mobile Crowd Sensing

Crowd sensing (CS) is a technique where a number of people having mobile devices can extract and share information related to their interests. CS sometimes is referred to as mobile crowd sensing (MCS) where smart devices such as phones and other wearable devices are emerging as sensing, computing, and communication devices [115]. MCS has a great potential of enhancing people's quality of life in various applications such as healthcare and transportation. Despite all the benefits of MCS, data privacy and user trustworthiness are critical problems which may face by it [115]. Therefore, these challenges are critical for CS in smart cities, which need to be carefully addressed in future research. Moreover, MCS has great environmental and social applications for smart cities. For example, environmental applications of MCS include measuring the level of pollution in a city and the level of water in creeks, and monitoring the natural home of wildlife [116]. Furthermore, in social MCS, users can share their sensed information using a database server, which provides a good understanding of problems related to the community [116].

12.8.2 Big Data

The rapid rise in the number of smart devices and big data leads towards the problem of security and privacy in smart cities' applications such as intelligent systems. Attackers can abuse the human intelligence and gain access to the big data in order to infer and violate the privacy of data owners. There are many techniques such as cryptographic approaches that can be used to detect these attackers. Besides that, it would be beneficial to improve the traceability of the network and allow a TTP to control it. In addition, it is of great importance to achieve data privacy, integrity, authentication, and availability in order to secure big data.

12.8.3 IoT-Based Network Security

The IoT can be considered as HetNets where different networks such as smart-phone networks, social networks, Internet, and industrial networks are integrated and connected together to provide better services to the people [117]. Due to this complex environment, it is required to conduct research on effective technologies in order to overcome the latest challenges regarding security and privacy in smart cities. In this regard, development of effective prevention approaches is significantly important. Moreover, it is beneficial to model the spread patterns of data in WSNs [118].

12.8.4 Lightweight Security Solutions

Several approaches have been developed recently regarding the security and privacy issues in smart cities. However, the application for some of these approaches is not realistic. Due to the availability of resource- and energy-constraint devices (e.g., sensors) in smart cities' infrastructures, it is not possible to implement advanced and strong security algorithms. Therefore, it is required to conduct research on developing lightweight security solutions to reduce overhead while providing an acceptable level of protection simultaneously.

12.8.5 Authentication and Confidentiality

In smart systems, authentication is required to ensure that services in heterogeneous systems can only be accessed by authorized users [119]. IoT devices in smart cities are capable of authenticating the network itself and other nodes in the network as well as the messages from the management stations. In addition, since the amount of data authentication is increasing dramatically, it is crucial to develop effective and advanced technologies to ensure exact and real-time authentication in smart cities.

Confidentiality is another requirement for securing smart cities. It prevents information from being subjected to the wrong source or passive attacks. In IoT networks, attackers can access devices and listen secretly to the communication. Therefore, it is important to conduct research into encryption-based techniques in order to protect confidentiality of data transmission between nodes. This in turn helps to have a reliable communication system [120].

12.8.6 Availability and Integrity

In smart cities, services should be available whenever they are needed. They should also be capable of maintaining effective actions while they are under attack. Furthermore, a smart system in smart cities should have the ability to discover unusual conditions and must be able to stop more damages to the system. Therefore, it is required to investigate on robust protection techniques to deal with the increasingly smart attacks.

Moreover, integrity of IoT devices and information being exchanged between the devices and the cloud is significantly important. Communications occur between various devices in smart cities; therefore, it is possible to easily damage the data if they are not properly protected during the transmission process. Thus, it is important to investigate on effective methods to guarantee the integrity of data in communication between IoT devices in smart cities.

12.8.7 The Application of Cloud/Fog Technology in Smart Cities

Cloud computing provides a way to locally store and manage an enormous amount of data collected by IoT devices. However, the process of sending such a vast amount of data is costly in terms of storage, bandwidth, latency, and communication. As a result, IBM suggested to process data at the edge of the network using the concept of fog computing, instead of transferring such a huge amount of data to the cloud [121]. Fog computing has several applications in achieving the requirements for building sustainable smart cities such as smart agriculture [122], smart healthcare [123], and smart water management [124]. In spite of the important benefits of fog computing for smart cities, several security and privacy challenges need to be taken into account such as data/web security and virtualization [125]. This is due to the limitation of computing resources of fog nodes, which complicates providing security solutions for them [126]. Moreover, the probability of cyber-attacks against fog nodes is higher than cloud data centers because they are usually more accessible.

12.9 CONCLUSION

Smart cities can improve the functionality of urban environments and enhance the quality of life and the well-being of people. By the implementation of various smart systems, security and privacy challenges have become an important issue which needs efficient and effective solutions. In addition, it is of great importance to consider the security and privacy threats in the design and implementation of new smart systems. In this chapter, we have investigated and discussed the security and privacy issues in information-centric smart cities' applications. First, we have introduced some typical applications of smart cities. We then presented the general requirements regarding the security and privacy challenges for smart cities' services. Moreover, we presented a number of security and privacy solutions for various applications of information-centric smart cities. We finally discussed some open research issues that should be carefully taken into account regarding the performance improvement of smart cities in terms of security and privacy.

REFERENCES

1. L. Cui, G. Xie, Y. Qu, L. Gao, and Y. Yang, "Security and privacy in smart cities: Challenges and opportunities", *IEEE Access*, 6, pp. 46134–46145, 2018.
2. E. Ever, F.M. Al-Turjman, H. Zahmatkesh, and M. Riza, "Modelling green HetNets in dynamic ultra-large-scale applications: A case-study for femtocells in smart-cities", *Computer Networks*, 128, pp. 78–93, 2017.
3. Y. Li, Y. Lin, and S. Geertman, "The development of smart cities in China", In *Proceedings of the 14th International Conference on Computers in Urban Planning and Urban Management*, pp. 7–10, July 2015.
4. M. Wang, J. Wu, G. Li, J. Li, Q. Li, and S. Wang, "Toward mobility support for information-centric IoV in smart city using fog computing", In *IEEE International Conference on Smart Energy Grid Engineering (SEGE)*, pp. 357–361, August 2017.
5. Z. Sheng, S. Yang, Y. Yu, A.V. Vasilakos, J.A. McCann, and K.K. Leung, "A survey on the Ietf protocol suite for the Internet of Things: Standards, challenges, and opportunities", *IEEE Wireless Communications*, 20(6), pp. 91–98, 2013.
6. A.V. Vasilakos, Z. Li, G. Simon, and W. You, "Information centric network: Research challenges and opportunities", *Journal of Network and Computer Applications*, 52, pp. 1–10, 2015.
7. D. Mars, S.M. Gammar, A. Lahmadi, and L.A. Saidane, "Using Information Centric Networking in Internet of Things: A Survey", *Wireless Personal Communications*, pp. 1–17, 2019.
8. F. Al-Turjman, "Information-Centric Framework for the Internet of Things (IoT): Traffic Modelling & Optimization", *Elsevier Future Generation Computer Systems*, vol. 80, no. 1, pp. 63–75, 2017.
9. O. Hahm, E. Baccelli, T.C. Schmidt, M. Wahlisch, and C. Adjih, "A named data network approach to energy efficiency in IoT", In *IEEE Globecom Workshops (GC Wkshps)*, pp. 1–6, December 2016.
10. K. Zhang, J. Ni, K. Yang, X. Liang, J. Ren, and X.S. Shen, "Security and privacy in smart city applications: Challenges and solutions", *IEEE Communications Magazine*, 55(1), pp. 122–129, 2017.
11. K. Biswas, and V. Muthukkumarasamy, "Securing smart cities using blockchain technology", In *18th IEEE International Conference on High Performance Computing and Communications; 14th IEEE International Conference on Smart City; 2nd IEEE International Conference on Data Science and Systems (HPCC/SmartCity/DSS)*, pp. 1392–1393, December 2016.

12. M. Perevezentsev and F. &. Sullivan, "Strategic opportunity analysis of the global smart city market," *Frost & Sullivan, Tech. Rep.*, 2013.
13. V. Albino, U. Berardi, and R.M. Dangelico, "Smart cities: Definitions, dimensions, performance, and initiatives", *Journal of Urban Technology*, 22(1), pp. 3–21, 2015.
14. "Smart cities require smarter cybersecurity," Tech. Rep., 2016. [Online]. Available: http://www.govtech.com/
15. Y. He, F.R. Yu, N. Zhao, V.C. Leung, and H. Yin, "Software-defined networks with mobile edge computing and caching for smart cities: A big data deep reinforcement learning approach", *IEEE Communications Magazine*, 55(12), pp. 31–37, 2017.
16. Z. Maamar, T. Baker, M. Sellami, M. Asim, E. Ugljanin, and N. Faci, "Cloud vs edge: Who serves the Internet-of-Things better?", *Internet Technology Letters*, 1(5), p. e66, 2018.
17. R. Mahmud, R. Kotagiri, and R. Buyya, "Fog computing: A taxonomy, survey and future directions", In *Internet of Everything*, pp. 103–130, Springer, Singapore, 2018.
18. K. Liang, L. Zhao, X. Chu, and H.H. Chen, "An integrated architecture for software defined and virtualized radio access networks with fog computing", *IEEE Network*, 31(1), pp. 80–87, 2017.
19. N. Abbas, M. Asim, N. Tariq, T. Baker, and S. Abbas, "A mechanism for securing IoT-enabled applications at the fog layer", *Journal of Sensor and Actuator Networks*, 8(1), p. 16, 2019.
20. S. Cui, S. Belguith, P. De Alwis, M.R. Asghar, and G. Russello, "Collusion defender: Preserving subscribers' privacy in publish and subscribe systems", *IEEE Transactions on Dependable and Secure Computing*, 2019.
21. F.S. Ferraz, C. Sampaio, and C. Ferraz, "Towards a smart-city security architecture: Proposal and analysis of impact of major smart-city security issues", In *Proceedings of the First International Conference on Advances and Trends in Software Engineering (SOFTENG)*, pp. 108–114, April 2015.
22. T.A. Butt, and M.A. fzaal, "Security and privacy in smart cities: Issues and current solutions", In *Smart Technologies and Innovation for a Sustainable Future*, Springer, pp. 317–323, 2019.
23. A. Gharaibeh, M.A. Salahuddin, S.J. Hussini, A. Khreishah, I. Khalil, M. Guizani, and A. Al-Fuqaha, "Smart cities: A survey on data management, security, and enabling technologies", *IEEE Communications Surveys & Tutorials*, 19(4), pp. 2456–2501, 2017.
24. D. Eckhoff, and I. Wagner, "Privacy in the smart city—Applications, technologies, challenges, and solutions", *IEEE Communications Surveys & Tutorials*, 20(1), pp. 489–516, 2018.
25. S. Sicari, A. Rizzardi, L.A. Grieco, and A. Coen-Porisini, "Security, privacy and trust in Internet of Things: The road ahead", *Computer Networks*, 76, pp. 146–164, 2015.
26. L. Tan, and N. Wang, "Future internet: The Internet of Things" In *Third IEEE International Conference on Advanced Computer Theory and Engineering (ICACTE)*, vol. 5, pp. V5–376, August 2010.
27. I. Hwang, and D. Shin, "Application level network virtualization using selective connection", In *IEEE International Conference on Consumer Electronics (ICCE)*, pp. 1–2, January 2018.
28. K. Anoh, A. Ikpehai, D. Bajovic, O. Jogunola, B. Adebisi, D. Vukobratovic, and M. Hammoudeh, "Virtual microgrids: A management concept for peer-to-peer energy trading", In *Proceedings of the 2nd ACM International Conference on Future Networks and Distributed Systems*, p. 43, June 2018.
29. V.C. Gungor, D. Sahin, T. Kocak, S. Ergut, C. Buccella, C. Cecati, and G.P. Hancke, "Smart grid technologies: Communication technologies and standards", *IEEE Transactions on Industrial Informatics*, 7(4), pp. 529–539, 2011.
30. F. Al-Turjman, and D. Deebak, "Seamless Authentication: For IoT-Big Data Technologies in Smart Industrial Application Systems", *IEEE Transactions on Industrial Informatics*, 2020. DOI: 10.1109/TII.2020.2990741.

31. F. Al-Turjman, and M. Abujubbeh, "IoT-enabled Smart Grid via SM: An Overview", *Elsevier Future Generation Computer Systems*, vol. 96, no. 1, pp. 579–590, 2019.

32. J.O. Petinrin, and M. Shaaban, "Smart power grid: Technologies and applications", In *IEEE International Conference on Power and Energy (PECon)*, pp. 892–897, December 2012.

33. S.P. Mohanty, U. Choppali, and E. Kougianos, "Everything you wanted to know about smart cities: The Internet of Things is the backbone", *IEEE Consumer Electronics Magazine*, 5(3), pp. 60–70, 2016.

34. E.I. Vlahogianni, K. Kepaptsoglou, V. Tsetsos, and M.G. Karlaftis, "A real-time parking prediction system for smart cities", *Journal of Intelligent Transportation Systems*, 20(2), pp. 192–204, 2016.

35. A. Zanella, N. Bui, A. Castellani, L. Vangelista, and M. Zorzi, "Internet of Things for smart cities", *IEEE Internet of Things journal*, 1(1), pp. 22–32, 2014.

36. B. Tang, Z. Chen, G. Hefferman, T. Wei, H. He, and Q. Yang, "A hierarchical distributed fog computing architecture for big data analysis in smart cities", In *Proceedings of the ASE Big Data & Social Informatics*, p. 28, October 2015.

37. X. Li, R. Lu, X. Liang, X. Shen, J. Chen, and X. Lin, "Smart community: An Internet of Things application", *IEEE Communications Magazine*, 49(11), pp. 68–75, 2011.

38. D. Ding, M. Conti, and A. Solanas, "A smart health application and its related privacy issues", In *Smart City Security and Privacy Workshop (SCSP-W)*, pp. 1–5, April 2016.

39. J. Ni, X. Lin, K. Zhang, and X.S. Hen, "Privacy-preserving real-time navigation system using vehicular crowdsourcing", In *84th IEEE Vehicular Technology Conference (VTC-Fall)*, pp. 1–5, September 2016.

40. L. Catarinucci, D. De Donno, L. Mainetti, L. Palano, L. Patrono, M.L. Stefanizzi, and L. Tarricone, "An IoT-aware architecture for smart healthcare systems", *IEEE Internet of Things Journal*, 2(6), pp. 515–526, 2015.

41. K. Zhang, R. Lu, X. Liang, J. Qiao, and X.S. Shen, "PARK: A privacy-preserving aggregation scheme with adaptive key management for smart grid", In *IEEE/CIC International Conference on Communications in China (ICCC)*, pp. 236–241, August 2013.

42. S. Finster, and I. Baumgart, "Privacy-aware smart metering: A survey", *IEEE Communications Surveys & Tutorials*, 16(3), pp. 1732–1745, 2014.

43. T. Gong, H. Huang, P. Li, K. Zhang, and H. Jiang, "A medical healthcare system for privacy protection based on IoT", In *Seventh IEEE International Symposium on Parallel Architectures, Algorithms and Programming (PAAP)*, pp. 217–222, December 2015.

44. Z. Ning, F. Xia, N. Ullah, X. Kong, and X. Hu, "Vehicular social networks: Enabling smart mobility", *IEEE Communications Magazine*, 55(5), pp. 16–55, 2017.

45. M. Wagner, M. Kuba, and A. Oeder, "Smart grid cyber security: A German perspective", In *IEEE International Conference on Smart Grid Technology, Economics and Policies (SG-TEP)*, pp. 1–4, December 2012.

46. V. Delgado-Gomes, J.F. Martins, C. Lima, and P.N. Borza, "Smart grid security issues", In *9th IEEE International Conference on Compatibility and Power Electronics (CPE)*, pp. 534–538, June 2015.

47. S.T. Zargar, J. Joshi, and D. Tipper, "A survey of defense mechanisms against distributed denial of service (DDoS) flooding attacks", *IEEE Communications Surveys & Tutorials*, 15(4), pp. 2046–2069, 2013.

48. S. McGregory, "Preparing for the next DDoS attack", *Network Security*, 2013(5), pp. 5–6, 2013.

49. E. Logota, G. Mantas, J. Rodriguez, and H. Marques, "Analysis of the impact of denial of service attacks on centralized control in smart cities", In *International Wireless Internet Conference*, Springer, Cham, pp. 91–96, November 2014.

50. T. Brewster, "Smart or stupid: will our cities of the future be easier to hack", *The Guardian*, 2014.
51. A. AlDairi, "Cyber security attacks on smart cities and associated mobile technologies", *Procedia Computer Science*, 109, pp. 1086–1091, 2017.
52. L.B. Cédric, E. Darra, D. Bachlechner, M. Friedewald, T. Mitchener-Nissen, M. Lagazio, and K.U.N.G. Antonio, "Cyber security for smart cities-an architecture model for public transport", 2015.
53. S. Ijaz, M.A. Shah, A. Khan, and M. Ahmed, "Smart cities: A survey on security concerns", *International Journal of Advanced Computer Science and Applications*, 7(2), pp. 612–625, 2016.
54. K. Angrishi, "Turning Internet of Things (IoT) into internet of vulnerabilities (iov): IoT botnets", *arXiv preprint arXiv*:1702.03681, 2017.
55. C. Kolias, G. Kambourakis, A. Stavrou, and J. Voas, "DDoS in the IoT: Mirai and other botnets", *Computer*, 50(7), pp. 80–84, 2017.
56. I. Ghafir, V. Prenosil, M. Hammoudeh, T. Baker, S. Jabbar, S. Khalid, and S. Jaf, "Botdet: A system for real time botnet command and control traffic detection", *IEEE Access*, 6, pp. 38947–38958, 2018.
57. S. Chaudhry, H. Alhakami, A. Baz, and F. Al-Turjman, "Securing Demand Response Management: A Certificate based Authentication Scheme for Smart Grid Access Control", *IEEE Access*, vol. 8, no. 1, pp. 101235–101243, 2020.
58. R.H. Weber, "Internet of Things–New security and privacy challenges", *Computer Law & Security Review*, 26(1), pp. 23–30, 2010.
59. A.S. Elmaghraby, and M.M. Losavio, "Cyber security challenges in smart cities: Safety, security and privacy", *Journal of Advanced Research*, 5(4), pp. 491–497, 2014.
60. K. Christidis, and M. Devetsikiotis, "Blockchains and smart contracts for the Internet of Things", *IEEE Access*, 4, pp. 2292–2303, 2016.
61. A. Dorri, S.S. Kanhere, R. Jurdak, and P. Gauravaram, "Blockchain for IoT security and privacy: The case study of a smart home", In *IEEE International Conference on Pervasive Computing and Communications Workshops (PerCom Workshops)*, pp. 618–623, March 2017.
62. A. Lei, H. Cruickshank, Y. Cao, P. Asuquo, C.P.A. Ogah, and Z. Sun, "Blockchain-based dynamic key management for heterogeneous intelligent transportation systems", *IEEE Internet of Things Journal*, 4(6), pp. 1832–1843, 2017.
63. G. Zyskind, and O. Nathan, "Decentralizing privacy: Using blockchain to protect personal data", In *IEEE Security and Privacy Workshops*, pp. 180–184, May 2015.
64. A. Fu, S. Yu, Y. Zhang, H. Wang, and C. Huang, "NPP: A new privacy-aware public auditing scheme for cloud data sharing with group users", *IEEE Transactions on Big Data*, 2017.
65. R. Neisse, G. Steri, and I. Nai-Fovino, "A blockchain-based approach for data accountability and provenance tracking", In *Proceedings of the 12th ACM International Conference on Availability, Reliability and Security*, p. 14, August 2017.
66. M. Laurent, N. Kaaniche, Ch. Le, and M. Vander Plaetse, "A blockchain-based access control scheme, In *15th International Conference on Security and Cryptography*, pp. 168–176, July 2018.
67. Q. Jing, A.V. Vasilakos, J. Wan, J. Lu, and D. Qiu, "Security of the Internet of Things: perspectives and challenges", *Wireless Networks*, 20(8), pp. 2481–2501, 2014.
68. Z. Mahmood, H. Ning, and A. Ghafoor, "Lightweight two-level session key management for end user authentication in Internet of Things", In *IEEE International Conference on Internet of Things (iThings) and IEEE Green Computing and Communications (GreenCom) and IEEE Cyber, Physical and Social Computing (CPSCom) and IEEE Smart Data (SmartData)*, pp. 323–327, December 2016.

69. N. Li, D. Liu, and S. Nepal, "Lightweight mutual authentication for IoT and its applications", *IEEE Transactions on Sustainable Computing*, 2(4), pp. 359–370, 2017.

70. A. Abdallah, and X.S. Shen, "A lightweight lattice-based homomorphic privacy-preserving data aggregation scheme for smart grid", *IEEE Transactions on Smart Grid*, 9(1), pp. 396–405, 2018.

71. I. Jabbar, and S. Najim, "Using fully homomorphic encryption to secure cloud computing", *Internet of Things and Cloud Computing*, 4(2), pp. 13–18, 2016.

72. M.S.H. Talpur, M.Z.A. Bhuiyan, and G. Wang, "Shared–node IoT network architecture with ubiquitous homomorphic encryption for healthcare monitoring", *International Journal of Embedded Systems*, 7(1), pp. 43–54, 2014.

73. A. Alabdulatif, I. Khalil, A.R.M. Forkan, and M. Atiquzzaman, "Real-time Secure Health Surveillance for Smarter Health Communities", *IEEE Communications Magazine*, 57(1), pp. 122–129, 2019.

74. M. Mackintosh, G. Epiphaniou, H. Al-Khateeb, K. Burnham, P. Pillai, and M. Hammoudeh, "Preliminaries of orthogonal layered defence using functional and assurance controls in industrial control systems", *Journal of Sensor and Actuator Networks*, 8(1), p. 14, 2019.

75. J. Hao, C. Huang, J. Ni, H. Rong, M. Xian, and X.S. Shen, "Fine-grained data access control with attribute-hiding policy for cloud-based IoT", *Computer Networks*, 153, pp. 1–10, 2019.

76. H. Cui, R.H. Deng, G. Wu, and J. Lai, "An efficient and expressive ciphertext-policy attribute-based encryption scheme with partially hidden access structures", In *International Conference on Provable Security*, Springer, Cham, pp. 19–38, November 2016.

77. L. Zhou, C. Su, W. Chiu, and K.H. Yeh, "You think, therefore you are: Transparent authentication system with brainwave-oriented bio-features for IoT networks", *IEEE Transactions on Emerging Topics in Computing*, 2017.

78. R. Amin, R.S. Sherratt, D. Giri, S.H. Islam, and M.K. Khan, "A software agent enabled biometric security algorithm for secure file access in consumer storage devices", *IEEE Transactions on Consumer Electronics*, 63(1), pp. 53–61, 2017.

79. I. Natgunanathan, A. Mehmood, Y. Xiang, G. Beliakov, and J. Yearwood, "Protection of privacy in biometric data", *IEEE Access*, 4, pp. 880–892, 2016.

80. Y. Wang, J. Wan, J. Guo, Y.M. Cheung, and P.C. Yuen, "Inference-based similarity search in randomized Montgomery domains for privacy-preserving biometric identification", *IEEE Transactions on Pattern analysis and Machine Intelligence*, 40(7), pp. 1611–1624, 2018.

81. A. Singandhupe, H.M. La, and D. Feil-Seifer, "Reliable security algorithm for drones using individual characteristics from an EEG signal", *IEEE Access*, 6, pp. 22976–22986, 2018.

82. M.A. Alsheikh, S. Lin, D. Niyato, and H.P. Tan, "Machine learning in wireless sensor networks: Algorithms, strategies, and applications", *IEEE Communications Surveys & Tutorials*, 16(4), pp. 1996–2018, 2014.

83. X. Luo, D. Zhang, L.T. Yang, J. Liu, X. Chang, and H. Ning, "A kernel machine-based secure data sensing and fusion scheme in wireless sensor networks for the cyber-physical systems", *Future Generation Computer Systems*, 61, pp. 85–96, 2016.

84. M.E. Aminanto, R. Choi, H.C. Tanuwidjaja, P.D. Yoo, and K. Kim, "Deep abstraction and weighted feature selection for Wi-Fi impersonation detection", *IEEE Transactions on Information Forensics and Security*, 13(3), pp. 621–636, 2018.

85. S. Shamshirband, A. Patel, N.B. Anuar, M.L.M. Kiah, and A. Abraham, "Cooperative game theoretic approach using fuzzy Q-learning for detecting and preventing intrusions in wireless sensor networks", *Engineering Applications of Artificial Intelligence*, 32, pp. 228–241, 2014.

86. B. Biggio, P. Russu, L. Didaci, and F. Roli, "Adversarial biometric recognition: A review on biometric system security from the adversarial machine-learning perspective", *IEEE Signal Processing Magazine*, 32(5), pp. 31–41, 2015.

87. S. Miller, and C. Busby-Earle, "The role of machine learning in botnet detection", In *11th IEEE International Conference for Internet Technology and Secured Transactions (ICITST)*, pp. 359–364, December 2016.

88. C.W. Tsai, C.F. Lai, M.C. Chiang, and L.T. Yang, "Data mining for Internet of Things: A survey", *IEEE Communications Surveys & Tutorials,* 16(1), pp. 77–97, 2014.

89. K. Xing, C. Hu, J. Yu, X. Cheng, and F. Zhang, "Mutual privacy preserving $ k $-means clustering in social participatory sensing", *IEEE Transactions on Industrial Informatics*, 13(4), pp. 2066–2076, 2017.

90. L. Li, R. Lu, K.K.R. Choo, A. Datta, and J. Shao, "Privacy-preserving-outsourced association rule mining on vertically partitioned databases", *IEEE Transactions on Information Forensics and Security*, 11(8), pp. 1847–1861, 2016.

91. J. Saleem, M. Hammoudeh, U. Raza, B. Adebisi, and R. Ande, "IoT standardisation: challenges, perspectives and solution", In *Proceedings of the 2nd ACM International Conference on Future Networks and Distributed Systems*, p. 1, June 2018.

92. J.A. Jerkins, "Motivating a market or regulatory solution to IoT insecurity with the Mirai botnet code", In *7th IEEE Annual Computing and Communication Workshop and Conference (CCWC)*, pp. 1–5, January 2017.

93. J. Saleem, and M. Hammoudeh, "Defense methods against social engineering attacks", In *Computer and Network Security Essentials*, Springer, Cham, pp. 603–618, 2018.

94. A. Ståhlbröst, A. Padyab, A. Sällström, and D. Hollosi, "Design of smart city systems from a privacy perspective", *IADIS International Journal on WWW/Internet*, 13(1), pp. 1–16, 2015.

96. D. Preuveneers, and W. Joosen, "Privacy-enabled remote health monitoring applications for resource constrained wearable devices", In *Proceedings of the 31st Annual ACM Symposium on Applied Computing*, pp. 119–124, April 2016.

95. A. Cavoukian, *"Privacy by Design: The 7 Foundational Principles"*, Information and Privacy Commissioner of Ontario: Toronto, 5, 2009.

97. A. Kung, J.C. Freytag, and F. Kargl, "Privacy-by-design in its applications", In *IEEE International Symposium on a World of Wireless, Mobile and Multimedia Networks*, pp. 1–6, June 2011.

98. J. Jung, A. Sheth, B. Greenstein, D. Wetherall, G. Maganis, and T. Kohno, "Privacy oracle: A system for finding application leaks with black box differential testing", In *Proceedings of the 15th ACM Conference on Computer and Communications Security*, pp. 279–288, October 2008.

99. H. Choi, S. Chakraborty, Z.M. Charbiwala, and M.B. Srivastava, "Sensorsafe: A framework for privacy-preserving management of personal sensory information", In *Workshop on Secure Data Management*, Springer, Berlin, Heidelberg, pp. 85–100, September 2011.

100. M. Layouni, K. Verslype, M.T. Sandıkkaya, B. De Decker, and H. Vangheluwe, "Privacy-preserving telemonitoring for ehealth", In *IFIP Annual Conference on Data and Applications Security and Privacy*, Springer, Berlin, Heidelberg, pp. 95–110, July 2009.

101. S. Gürses, C. Troncoso, and C. Diaz, "Engineering privacy by design", *Computers, Privacy & Data Protection*, 14(3), p. 25, 2011.

102. D. Le Métayer, "Privacy by design: A formal framework for the analysis of architectural choices", In *Proceedings of the Third ACM Conference on Data and Application Security and Privacy*, pp. 95–104, February 2013.

103. A. Monreale, S. Rinzivillo, F. Pratesi, F. Giannotti, and D. Pedreschi, "Privacy-by-design in big data analytics and social mining", *EPJ Data Science*, 3(1), p. 10, 2014.

104. Q. Wang, K. Ren, S. Yu, and W. Lou, "Dependable and secure sensor data storage with dynamic integrity assurance", *ACM Transactions on Sensor Networks (TOSN)*, 8(1), p. 9, 2011.

105. K. Kursawe, G. Danezis, and M. Kohlweiss, "Privacy-friendly aggregation for the smart-grid", In *International Symposium on Privacy Enhancing Technologies Symposium*, Springer, Berlin, Heidelberg, pp. 175–191, July 2011.

106. J. Shi, R. Zhang, Y. Liu, and Y. Zhang, "Prisense: Privacy-preserving data aggregation in people-centric urban sensing systems" In *Proceedings IEEE INFOCOM*, pp. 1–9, March 2010.

107. T. Denning, C. Matuszek, K. Koscher, J.R. Smith, and T. Kohno, "A spotlight on security and privacy risks with future household robots: Attacks and lessons", In *Proceedings of the 11th ACM International Conference on Ubiquitous Computing*, pp. 105–114, September 2009.

108. S. Jha, L. Kruger, and V. Shmatikov, "Towards practical privacy for genomic computation", In *IEEE Symposium on Security and Privacy*, pp. 216–230, May 2008.

109. S. Chakrabarty, and D.W. Engels, "A secure IoT architecture for Smart Cities", In *13th IEEE annual consumer communications & networking conference (CCNC)*, pp. 812–813, January 2016.

110. S. Chakrabarty, D.W. Engels, and S. Thathapudi, "Black SDN for the Internet of Things", In *12th IEEE International Conference on Mobile Ad Hoc and Sensor Systems*, pp. 190–198, October 2015.

111. O. Flauzac, C. González, A. Hachani, and F. Nolot, "SDN based architecture for IoT and improvement of the security", In *29th IEEE International Conference on Advanced Information Networking and Applications Workshops*, pp. 688–693, March 2015.

112. F. Al-Turjman, "Intelligence and Security in Big 5G-oriented IoNT: An Overview", *Elsevier Future Generation Computer Systems*, vol. 102, no. 1, pp. 357–368, 2020.

113. D. Kelly, and M. Hammoudeh, "Optimisation of the public key encryption infrastructure for the Internet of Things", In *Proceedings of the 2nd ACM International Conference on Future Networks and Distributed Systems*, p. 45, June 2018.

114. V.L. Shivraj, M.A. Rajan, M. Singh, and P. Balamuralidhar, "One time password authentication scheme based on elliptic curves for Internet of Things (IoT)", In *5th IEEE National Symposium on Information Technology: Towards New Smart World (NSITNSW)*, pp. 1–6, February 2015.

115. D. He, S. Chan, and M. Guizani, "User privacy and data trustworthiness in mobile crowd sensing", *IEEE Wireless Communications*, 22(1), pp. 28–34, 2015.

116. P. Chithaluru, F. Al-Turjman, M. Kumar, and T. Stephan, "I-AREOR: An Energy-balanced Clustering Protocol for implementing Green IoT for smart cities", *Elsevier Sustainable Cities and Societies*, 2020. DOI: 10.1016/j.scs.2020.102254.

117. K. Xu, Y. Qu, and K. Yang, "A tutorial on the Internet of Things: From a heterogeneous network integration perspective", *IEEE Network*, 30(2), pp. 102–108, 2016.

118. S. Yu, G. Gu, A. Barnawi, S. Guo, and I. Stojmenovic, "Malware propagation in large-scale networks", *IEEE Transactions on Knowledge and Data Engineering*, 27(1), pp. 170–179, 2015.

119. D. He, S. Zeadally, N. Kumar, and J.H. Lee, "Anonymous authentication for wireless body area networks with provable security", *IEEE Systems Journal*, 11(4), pp. 2590–2601, 2017.

120. H. Zahmatkesh, and F. Al-Turjman, "Fog Computing for Sustainable Smart Cities in the IoT Era: Caching Techniques and Enabling Technologies - An Overview", *Elsevier Sustainable Cities and Societies*, vol. 59, 102139, 2020.

121. M. Sookhak, F.R. Yu, Y. He, H. Talebian, N.S. Safa, N. Zhao, M.K. Khan, and N. Kumar, "Fog vehicular computing: Augmentation of fog computing using vehicular cloud computing", *IEEE Vehicular Technology Magazine*, 12(3), pp. 55–64, 2017.

122. C. I. Centre, C. Sensor, and S. N. TCP, "Phenonet: Distributed sensor network for phenomics supported by high resolution plant phenomics centre", *Commonwealth Scientific and Industrial Research Organisation (CSIRO), Tech. Rep.*, 2011.

123. J. K. Zao, T. T. Gan, C. K. You, S. J. R. Mndez, C. E. Chung, Y. T. Wang, T. Mullen, and T. P. Jung, "Augmented brain computer interaction based on fog computing and linked data", In *International Conference on Intelligent Environments*, pp. 374–377, June 2014.

124. C. Perera, Y. Qin, J. C. Estrella, S. Reiff-Marganiec, and A. V. Vasilakos, "Fog computing for sustainable smart cities: A survey", *ACM Computing Surveys*, 50(3), p. 32, June 2017.

125. M. Sookhak, R. Yu, and A. Zomaya, "Auditing big data storage in cloud computing using divide and conquer tables", *IEEE Transaction on Parallel and Distributed Systems*, 29(5), pp. 999–1012, 2018.

126. P. Zhang, J. K. Liu, F. R. Yu, M. Sookhak, M. H. Au, and X. Luo, "A survey on access control in fog computing", *IEEE Communication Magazine*, 56(2), pp. 144–149, 2018.

Index